A pract… tting
… of grassland

Grassland signals

Berrie Klein Swormink | Nick van Eekeren | Bert Philipsen

ROODBONT
UITGEVERIJ
PUBLISHERS

LIVESTOCK RESEARCH
WAGENINGEN UR

LOUIS BOLK
INSTITUUT

Straight Line Nutrition Ltd.

Credits

Grassland signals

A practical guide for getting the most out of grassland

Authors

Nick van Eekeren, Louis Bolk Institute
Bert Philipsen, Animal Sciences Group, Wageningen University
Berrie Klein Swormink, Klein Swormink Rural Business

Translation from Dutch

Noorderlicht

Editor English edition

Luppo Diepenbroek

Thanks to:

Robert Fakkert, Jan Hollander, Henk Jansen, Jos de Kleijne, Wichert Koopman, Praktijkcentrum Aver Heino, Henk Schilder, René Verhoeven, Jan Visscher, Marcel Vijn, Lex van der Weerd, Jan de Wit

Photography

Berrie Klein Swormink, Louis Bolk Instituut, Animal Sciences Group; also: Agrifirm (94), Marcel Batten (78 r), Marcel Bekken (75 br), Blgg (16) , Dreamstime (33 br, 72 c and r, 73 l), Geometius (75 l), Google Earth (91 br), Grasdrogerij Hartog (91 tr), Silvia Hellingman (73 tr), Rowie Klein Swormink (19, 61 c, 93 bl), LaMi (30 tl), Lely (61l), Twan Wiermans (10 br, 24 br, 36 bl, 44 br), Wikipedia Commons, user 'Rasbak' (81 tc, tr and br); Cover photo: Twan Wiermans
t = top, b = bottom, l = left, r = right, c = centre

Illustrations

Gerda Peters

Design

bhgo, design for web and print

LIVESTOCK RESEARCH WAGENINGEN UR

P.O. box 65
NL-8200 AB Lelystad
The Netherlands
T +31(0)320 23 82 38
F +31(0)320 23 80 50
E info.livestockresearch@wur.nl
I www.livestockresearch.wur.nl/uk

LOUIS BOLK INSTITUUT

Hoofdstraat 24
NL-3972 LA Driebergen
The Netherlands
T +31(0)343 52 38 60
F +31(0)343 51 56 11
E info@louisbolk.nl
I www.louisbolk.nl

KleinSwormink

Assinksteeg 1
NL-7434 PR Lettele
The Netherlands
T +31(0)570 54 14 96
F +31(0)570 54 32 73
E info@kleinswormink.nl
I www.kleinswormink.nl

ROODBONT UITGEVERIJ PUBLISHERS

P.O. Box 4103
NL-7200 BC Zutphen
The Netherlands
T +31(0)575 54 56 88
F +31(0)575 54 69 90
E info@roodbont.nl
I www.roodbont.nl

Straight Line Nutrition Ltd.

Loosedon Barton Bungalow
Winkleigh
Devon EX19 8HA
England
T 01837 680000
E luppod@o2.co.uk
I www.watchmygrass.co.uk

ISBN: 978-90-8740-046-0

Table of Contents

Grassland signals

On livestock farms, grass is usually the main source of feed for the animals. Good grassland production hugely influences the business's economic performance. Taking an interest in grassland can provide good returns.

Grassland production varies widely, ranging from about 6 to 16 tonnes of dry matter per hectare per year. The differences can be partly explained by variations in natural conditions such as soil type, soil fertility and water supply. However, with good grassland management, you can also influence the yield and quality of the grass. The economic results of a dairy farm are closely related to grassland yield.

The good thing about grassland is that it provides many signals that you can use to fine tune its management. The grassland make up is a strong indicator about growing conditions and use. Signal plants, for example, can help you take management decisions. They give an indication of the conditions in which they thrive – for example the presence of couch grass can indicate poor soil structure.

Other signals, which are often disregarded, can be found below the surface. The soil below grassland is of crucial importance for good production. In good soil, the soil particles, air and water are randomly, but intimately dispersed. It's teeming with life and roots have space to grow and utilise the nutrients they need. Taking a regular close look at your soil should be a frequent exercise in your grassland management.

Sometimes grassland needs to be treated a bit rough. But you have to know the limits. ▼

Grazing

Grassland signals is not a grassland handbook. You won't find elements of grassland management arranged by subject; the arrangement is based on the situation on grassland throughout the year. Some activities, like grazing, cutting or tackling weeds, take place at several points throughout the year. You'll come across these activities on several occasions in this book too, always from a theme that's important at that time of year.

Use this guide as a reference. Translate the information in the text and illustrations into your own grassland practice, and look at it as a challenge to observe grassland in a new perspective picking up new signals. The handy index at the back will make it easier to 'graze' through *Grassland signals.*

Observational book

Grassland signals is not a book to be read from front to back at one sitting. It's about observing and reflecting. Text, but above all the many photos and illustrations make it clear what can be seen in and below grassland. They'll help you to see the grassland from a different perspective.

Puzzler

Puzzler

Here and there in this book you'll find 'Puzzlers' – pictures that playfully challenge you to take a closer look at the situation pictured. Looking more conscientiously will give you information you might otherwise overlook. The Puzzlers will help you to think about what you see.

Chapter 1

Grass is the basis

Grasses have adapted themselves to grazing over the ages. On the other hand, grazing animals have become better equipped over time at eating and digesting grass. Knowing how grasses and other grassland plants grow is useful for good pasture management and optimum grassland composition.

▲ *In the course of its evolution, grassland has adapted to short, intensive periods of grazing. Bison, for example, come in as a herd, eat the grass short and move onto the next piece.*

Grassland serves first and foremost to produce food for ruminants (cattle, sheep and goats). In nature, most ruminants live in herds to protect themselves from predators. They can eat large quantities of grass at a time, and ruminate it later in a safe place. Under natural conditions a herd will graze an area close to the ground and move on. The current system of strip-grazing or short-rotation grazing most resembles their natural pattern: short and intensive. The shape of their jaws, tongues and lips enables ruminants to eat vegetation close to the ground. Grasses have adapted themselves superbly to this kind of grazing behaviour.

Grazing cows for a long time on grass shorter than 10 cm is unnatural and comes at the expense of growth (for example under set stocking). The plants have too little leaf surface to supplement the food reserves in their roots. The root systems develop less well, making the grass less resistant to drought and stress. ▼

Two qualities of grass prove that it has adapted to grazing:

- The growing point is low to the ground, so it is not easily damaged by grazing. Most other plants have the growing point at the top of the plant.
- Energy can be stored in the roots, and used later for re-growth.

How grass grows

After grass has been grazed or cut it starts to grow after a few days, using the energy in its root reserves. Growth is at the expense of the root system, which partially dies back.

During the first week of this growth, the young leaf only has a small surface area, and it can absorb very little energy. Thus the first growth is slow and is fuelled mainly from the root reserves. As the leaf surface increases, the plant can collect more and more energy from the sun. The speed of growth increases and reaches a maximum, until the entire soil surface is covered with grass blades. This is generally the case if there is a grass dry matter yield of 1200 kg (2600 kg DM herbage mass (presence)) per hectare of land.

At a certain point, the plant is using so much solar energy that it is taking up more energy than it can use for leaf and stem growth above the soil surface. From that point on, it diverts energy to the roots, thereby continually developing a new balance between above-ground and below-ground biomass.

◄ *Grass has its growing points low to the ground. They are not easily eaten during grazing and the plant can re-grow within a few days.*

'You need grass to grow grass' is an old saying. As the surface area of a grass leaf increases, so does its rate of growth. It is at its greatest when it reaches its optimal 'leaf area index', or when every ray of sunlight ends up on a grass blade. ▼

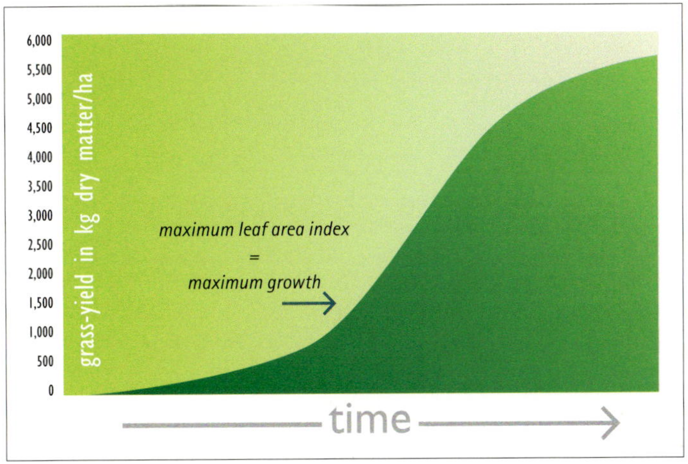

Grass grows to death

Grass plants get into trouble if they are not cut or grazed. A too tall or too dense crop will not allow enough sunlight to the young grass tillers down below, and these tillers will die. After flowering and forming seed-heads, the main tillers will also die. The sward will try to maintain itself by forming new tillers. There is no energy to replenish root reserves and the grass will start to die from the centre outwards. This phenomenon used to be seen often in conservation areas in Europe which were not grazed.

Grass perfume

Cis-3-hexene-1-ol is an organic compound responsible for the smell of freshly mown grass. This 'leaf alcohol' is a colourless, oily liquid, and occurs in various green plants. This 'grass perfume' also produces the taste of green tea.

Grasses

Worldwide, there are more than 8000 species of grasses. The smallest are only a few centimetres high, and the largest can reach about six metres. Livestock farmers in the UK and the Netherlands only have to deal with a few dozen species. The strength of grass is in numbers – grass plants are capable of crowding out other plants by forming a dense and extended carpet of vegetation. Grasses can reproduce both vegetatively and by forming flowers from which seeds are produced. Vegetative or asexual reproduction takes place sometimes through formation of tillers and stolons, from which new plants grow. Mowing, but above all grazing, promotes vegetative reproduction. The grass sward is often denser where livestock are grazing.

Spring treatment with a grass harrow will remove old leafy matter and introduce air into the soil. ▼

Grasses and grazers cannot survive without each other. Grazers eat not only the grass but other plants as well, which would otherwise grow bigger than the grass and block their sunlight supply. And of course, grazers fertilise the grass too. ▼

Although cereals like triticale and maize are annuals, they also belong to the grass family. ▼

Clover for nitrogen

Grassland contains many other plants besides grasses. Sometimes these are species that are deliberately introduced into grassland, for example clovers. But there can also be less desirable species – the ones we usually call weeds.

Clover complements grass very well, because like other legumes it uses bacteria in its root nodules to fix atmospheric nitrogen. Grass benefits from the nitrogenous compounds produced by the clover to grow and produce plant protein.

As early as the fifteenth and sixteenth centuries, farmers were experienced in introducing clover into their sward. However, cultivating on poorer soils with low fertility was not so easy in those days. After artificial fertiliser was introduced in the late nineteenth century, interest in clover waned. At the end of the twentieth century it was mostly organic farmers who kept on working with clover in their grasslands. Because nitrogen applications are limited by legislation and fertilizer costs are continuing to rise, interest in clover is increasing again.

▲ *There are several varieties of white clover: small-leaved (suitable for grazing), medium-leaved (suitable for grazing and cutting) and large-leaved (suitable for cutting, but also suitable in cutting-grazing systems).*

▲ *Grassland with red clover can yield more than 15 tonnes of dry matter per hectare per year without the use of nitrogen fertilisers.*

Grassland usually contains plants other than those which were sown. For example, most livestock farmers are not happy to see docks appear in their pastures. ▼

With good grassland management you can influence the proportions of grass to clover. ▼

◀ *Sheep sorrel contains oxalic acid, which can cause diarrhoea in cattle and sheep. Eating a lot of oxalic acid can block an animal's calcium uptake. Sheep sorrel is an indicator for a low pH.*

◀ *Red clover contains phyto-oestrogen, which is similar to the animal hormone oestrogen. Grazing red clover with ewes during the breeding period can reduce fertility.*

Changing growth

The growth of grass changes quite a lot throughout the year, making it important to follow its development. After a winter rest, the grass starts to grow again once the soil warms up enough. Under ideal conditions, grass growth will mostly depend on light and temperature. In the spring there is a sharp increase in the growth rate, which reaches a peak in May. Thereafter it gradually slows down, virtually stopping in October-November. In practice, there often seems to be a dip in growth in June-July (see graph). More than one factor seems to be at work: heavy first and second silage cuts delaying re-growth, botanical composition, lack of moisture or minerals and high temperatures. The point at which a summer decline occurs can thus also vary from year to year.

The 'camel chart' (bi-modal chart) is a casual term for the representation of grass growth over the year. There is often a dip in June/July, making what looks like two 'humps'. The light green line shows grass growth in a year with a warm spring and a wet summer. The dark green line shows a year with a cold spring and a dry summer. ▼

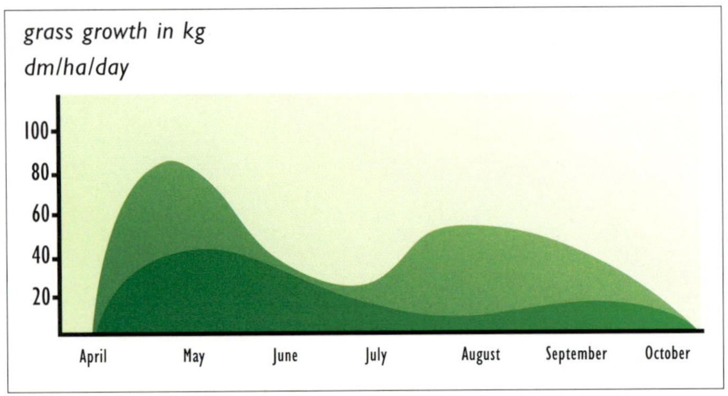

Perennial ryegrass is the most commonly used grass species. It produces high yields, responds well to nitrogen fertilisers and grows well in our climate. It is also readily eaten by grazing animals. ▼

Black-coloured red clover leaves in the silo are a good sign. The black comes from the enzyme polyphenol oxidase (PPO) in red clover. A positive effect of this enzyme is that it protects proteins in the silage from breaking down. PPO occurs in other areas of daily life, for example, causing the black colouring in bananas. ▼

Feed the cycle

Grass, root growth, soil organism and soil together form a cycle. Roots grow and die back again. This is a continuous process in which much organic matter ends up in the soil. The bacteria and other organisms in the soil convert this material into nutrients for plants and create room for the roots to grow. A good root layer can grow 4500 kg/hectare of root mass.

A turf layer with a good root system is the most important tool a livestock farmer has to improve the soil. Grasses have many fine roots which penetrate everywhere.
So, stimulating root growth has a positive effect on the soil organisms and soil structure, which in turn promotes further root growth and better grass growth above as well as below ground.

Work with machines that compact the soil as little as possible. A low tyre pressure is important. Using a pressure exchange system like the one on this self-driving grassland injector allows you to drop your tyre pressure on the grassland as far down as possible. ▼

Grass, root growth, soil organisms and soil form a self-reinforcing cycle. Ensuring good root growth from the time of sowing is one of the most important elements of pasture management. It will keep the flywheel behind the cycle turning. ▼

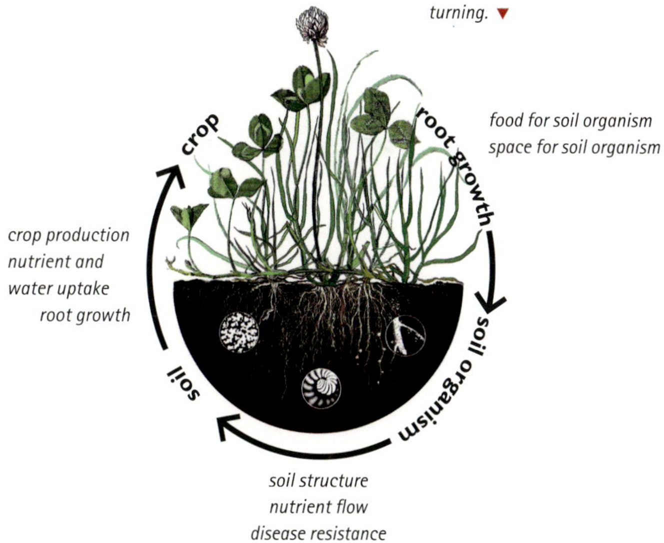

Fresh grass makes healthier milk

Grass influences the composition of the milk produced by the cows that eat it. Cows that eat a lot of fresh grass produce milk with a large number of healthy unsaturated fatty acids, also known as CLAs (or conjugated linoleic acids). These are mostly omega-3 fatty acids. In particular linoleic acid, which is present in large amounts in young grass, seems to be good for milk quality. It turns out that cows allowed to graze produce healthier milk than cows being served a summer diet indoors. This is because grazing cows are mostly eating the tips of grass plants, where most linoleic acid is found.

Chapter 2 December-January

Winter rest

▲ *Grassland is vulnerable in the winter.*

Growth slows down during the winter months. It's important that the grassland survives through the winter and receives a good start in spring. Take a look now at the quantity and quality of manure available for the sward. Now is the time to think about what part the grassland plays in your cropping plans for the new season.

Care with sheep

Grass that begins the winter too long is vulnerable to winter damage. This can be from frost damage or from suffocation from snowfall, for example. If the grass is longer than 8–10 cm, sheep grazing can help. Sheep are also good for managing clover in pastures. However, too intensive grazing by sheep will be at the expense of re-growth in the spring. Don't let the grass get shorter than 4-5 cm, and avoid poaching it. Make sure they're off the pasture before the end of January. This will prevent growth from being inhibited once the plants awake from their winter sleep. Moreover, grazing causes damage in the event of an extended frost.

There's a saying that 'sheep have golden hooves'. Many cattle farmers let sheep graze their pastures in autumn and winter. They weigh less and thus do less damage, even in winter when the grass sward carrying capacity is reduced. ▼

Sheep shouldn't graze on young grass. Newly seeded grassland is highly vulnerable to poaching and compaction of the topsoil.

Moles

Combating moles requires continuous attention in the winter. Their digging damages the grass sward creating opportunities for weeds. Molehills also raise the ash content of the silage.

Getting and keeping your grassland mole-free is best done by joining forces with your neighbours. Otherwise, once you're rid of your moles, the neighbouring moles will quickly move in to the newly available territory.

It is a good sign if moles like a particular grass field, as it indicates that a lot of worms and other soil organisms are present.

The most effective and environmentally-friendly way to catch moles is with a clamp trap, although correct placement of the traps requires a bit of practice. Try to catch moles at the edges of the field by placing traps in mole runs (tunnels). Set a marker at each trap site so you can find it later. ▼

Leatherjackets – problems above water

Leatherjackets are the larvae of the crane fly (daddy long-legs). They damage the root layer by eating the above-ground parts of the grass plants. They draw grass blades downwards from ground level. In general, chemical pesticides only pay off if there are more than 300 leatherjackets per square metre. Do the first leatherjacket count in December.

Leatherjackets can be counted easily by submerging a piece of turf 10 cm long, 10 cm wide and 5 cm thick in a salt bath (1 kg salt to 5 litres water). They will rise to the top of their own accord, and are easy to count. Multiply the number by 100 and you'll have the number present per square metre of grassland.

▲ *The soil's carrying capacity will be greatly reduced during a wet spring if you have poor drainage. Allowing machinery on the land too early can quickly cause damage to the grass sward.*

Drainage

Good drainage will help you maintain the desired plant composition of your pasture. Puddles should disappear within 24 hours after heavy rainfall. If water stands any longer, then the root zone will start to lose oxygen, affecting the soil organism and the grass roots. This in turn makes badly drained grassland more vulnerable to dry weather in summer.

Good drainage ensures that the soil has good carrying capacity, allowing manure to be applied to the grass earlier in the spring. This works out well, because well-drained fields warm up faster in the spring, allowing the grass to make an earlier start than in wetter fields. Driving machinery onto waterlogged grassland causes structural damage and leads to compaction of the subsoil, making the drainage of the field even worse.

Grassland is considered to have good drainage if the average groundwater level in the winter months is more than 40 cm below the soil surface. Try to get a picture of groundwater levels by drilling holes and measuring the height of the groundwater.

Combating moles is desirable because they destroy the root layer and get soil in the crop. However, from the drainage point of view, having moles in the ground is not a bad idea. ▼

Digging a few small trenches can help remove excess water off the land after long or heavy rain. This can make the difference between keeping and losing the grass sward, especially on new grassland. The need to dig trenches is, however, a sign that the parcel is not entirely level. You should consider levelling on your next round of improvements. ▶

Measures to improve drainage include:

- digging ditches to improve water flow;
- lowering ditch-water levels in the winter season and during heavy rainfall during the growing season;
- levelling soil surfaces and grading within fields;
- installing drains;
- digging furrows.

What measure you use will depend on the region and soil type. Economic considerations also play a part: weigh up any increased profits against the costs. Water levels may also be managed by local water boards.

▲ *A wet area will make itself worse. The lack of oxygen will cause a decrease in growth and the amount of roots, and thereby the permeability of the soil, around a wet area. The soil compacts and subsides. A wet area can expand this way from year to year.*

Digging small furrows can help surface drainage on wet fields. Don't make wide, deep channels as they will result in low spots in the field. Don't roto-trench old channels if there is standing water in them. ▼

Checking drainage pipes

Drainage can really help remove water. Winter is an excellent time to check that everything is working well on drained fields. Check whether water is flowing out of the end pipes and that there is no sludge blocking them. High-pressure cleaning can be a good maintenance measure if the drainage pipes are not running enough water and seem to be fouled.

There are scientific formulae for calculating whether the run-off from the drains is still sufficient. Observation in the field is more practical: is the grassland wetter than it used to be? Do you see the drains flowing freely after a period of rainfall?

Soil analysis – worth doing

Soil fertility determines to a large extent the productivity of grassland and the quality of the grass grown. Regular soil analysis can help you to fine-tune your manuring to the current state of the soil. Livestock can make use of the soil analysis, which they have to produce, in order to comply with catchment directives, or in NVZs, to determine soil phosphorus, potassium, and magnesium, with pH. Tip: When taking soil samples for any purpose, perform a more comprehensive soil analysis at the same time.

Organic matter (OM)

The amount of organic matter speaks volumes about soil quality. Grassland should have a rather high OM content. Organic matter, or humus, allows minerals to stay in the soil and not be leached. Humus can also absorb moisture, making grassland with soil high in OM less vulnerable during dry periods. Every additional percentage point of organic matter provides six millimetres more moisture. Adequate OM content also prevents soil compacting. Every year, grassland produces a large amount of organic matter in the form of roots and crop remains, which finish up in the soil.

Organic matter is also delivered in the form of manure. With permanent grassland, there is an equilibrium between organic matter deposited and removed. Ploughing up the grassland disrupts this balance, causing a drop in the OM after re-seeding, so you should 'renew' grassland as little as possible. After re-seeding, the OM will take years to come back to previous levels. The OM content can differ sharply by soil type and region: peaty soil contains more than 25 percent organic matter. A sward on sandy soil is good if it's 6 percent.

Soil analysis provides information about the nutritional situation. However, what you feed your grass and clover is not the only important aspect; the mineral composition for the animals health plays a role as well. For example, it's desirable to have enough cobalt in the soil: between 8 and 18 micrograms per kilogram of soil. Cobalt in the grass is important for rumen function. ▼

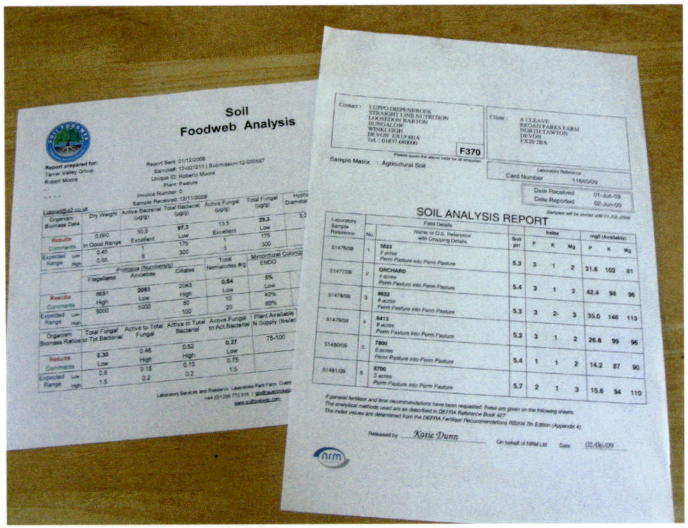

Soil samples from a given field must be representative of the soil under the entire field area. If a field is not homogeneous in soil and grass production, it's better to sub-divide the field; chemical analyses will then give you more useful information. ▼

Soil Nitrogen Supply (SNS)

The SNS Index system represents the soil's ability to supply nitrogen: the nitrogen not provided by that year's fertiliser that becomes available in the soil for the plant. On grassland, the SNS contributes quite a lot to supplying nitrogen for the grass plants. A soil with high SNS needs less nitrogen from fertiliser. There is a strong connection between the SNS and the OM content of the soil. A rise of one percent in OM means roughly an increase in SNS by 25 kg N per ha per year.

Acidity (pH)

Many farms experience a drop in soil pH (measure of acidity) over the years due to fertiliser use and precipitation of acidifying matter. The soil becomes more acid. With grass swards the pH must be at least 5, but no higher than 6. The optimum pH for grassland containing clover is 5.5. Clover plants fix less nitrogen at a pH lower than 5.2, which is the highest pH many peaty soils can achieve.

Farm yard manure is high in phosphates

Farm yard manure contains relatively high levels of phosphates. Soil plays an important role in supplying phosphate. Phosphate is much less mobile than nitrogen. Highly productive grassland consumes about 100 kg phosphate per ha per year. Aim for maintenance dressing if there is enough phosphate in the soil. Active soil organisms help phosphate in the soil to become more readily available. Generally, the soil below every hectare of grassland contains 2.5 to 5 tonnes of phosphate, which is not immediately available but can become available. Soil organisms and roots also play an important part in this process.

◄ *The K Index indicates how much potash is in the soil. Stricter standards for the use of animal manures mean that potash supply can be limited, especially where silage is cut. The development of the available potash can be read from soil analyses over several consecutive years. Clover suffers earlier than grass when the potash content is low.*

Choosing fertiliser

A manure plan provides clarity about the availability and distribution of manure. It's also clear how much room government regulations allow for applying nitrogen and phosphate in the form of fertilisers. For grassland, the use of fertiliser is mainly focused on nitrogen. If other elements are also needed, such as phosphate, sulphur, magnesium or trace elements, these can be combined when you choose fertiliser types or blends. Bear in mind that fertiliser input can also mean undesirable heavy metals, such as copper in kieserite.

Plan your manuring

During the winter months, make a plan for manuring during the next growing season. Under farm assurance, catchment regulations and NVZ rules you must have prepared a manure plan every year. The basis of the manure plan is the amount of available nutrients from farm yard wastes. Take annual samples of slurry and farm yard manure, as it gives an idea of their nutrient value. You can also estimate quite accurately in total how much nitrogen and phosphate is available from animal manures.

First of all, calculate the amount of manure available for the total farm surface area. If there is more manure than is legally permitted on the area in use, some of the manure must be removed. On extensive holdings, it is sometimes possible to bring in manure from outside.

Look back at previous season

See whether previous season's manure plan performed as intended. Did it result in the anticipated grassland yields? Compare the estimated annual yield with the total fertiliser applied for each field. Check whether the grass yield corresponds to the yield for which it was fertilised. If the yield for a given field is less than anticipated year on year, despite having enough fertiliser, you can lower the amount of fertiliser on it for the coming growing season. This will give you room to increase fertiliser application on the more vigorous fields. When applying within rows or when growing maize on ploughed-up grassland the slurry application can be decreased more.

Distribution

If you are growing other crops besides grass, it is important to distribute manure evenly over the various crops, especially if the regulations do not allow all crops to be optimally fertilised. Livestock farms may have maize in addition to grass. From a mineral uptake and nutritional yield standpoint, it is desirable to cut down the manure use on grassland in favour of maize fields. A practical approach is to adjust the slurry applied to maize fields to the maize's phosphate requirements. This often means that about 35 m^3 of cattle slurry per hectare will be used on maize fields.

Distribute the remaining manure over the grassland. How much you can spread on the grassland will depend to a significant degree on your grazing system. Unlimited grazing by milking cows and calves means you'll have to stockpile less manure. RB 209 Fertiliser recommendations indicate what nutrients grassland needs for optimum production.

Usually the maximum levels will be limited by government standards for use. Using the standards for the crop and for manure, you can calculate how much nitrogen and phosphate from fertilisers may be applied to the land in addition to animal manures.

▲ *If you use different kinds of manure, working out the manure plan will be a bit more complicated. For example, about 50 percent of the nitrogen from cattle slurry is available for the grass in the year it is applied. By contrast, no more than 20 percent of the nitrogen in cattle farm yard manure is available, with the remainder being freed up in the following years.*

◂ *Analysing the available manure will not only provide information about the manure's nutrient value, but will tell you about the animals' nutrition. A phosphate content of less than 1.3 kg per tonne (based on 9% dry matter) in cattle manure can indicate a lack of phosphate in the cows' diet.*

In areas such as peat meadows, where the ground is only suitable for grass, coming up with the optimum use plan is not difficult. ▶

Bulb cultivation goes well in rotation with grassland. After growing bulbs the land is well suited to sowing grass/clover. The soil contains little nitrogen, which favours the establishment of clover, and you save a lot on nitrogen fertiliser. ▶

If you need to improve the grassland, for example because (as shown here) there are too many weeds or too few good grasses, one good idea may be to grow another fodder crop. The crop will shift the re-seeding of the grassland to the autumn, and production loss from re-seeding will be less. ▶

Long term or temporary?

Grasslands can be divided into long term and temporary grassland. Long term grassland is used for several consecutive years as pasture. With temporary grassland (one or two years), grass is part of a rotation plan that includes other crops. The decision whether to rotate or opt for long term grassland, possibly combined with other crops on another part of the farm, depends on many factors. The farming situation, soil type and farm layout all play a large part. Some soil types such as peaty soils are good for little else but grassland. In these situations grass is usually the only crop grown and the best choice will be for long term grassland. On other soils, such as sand, clay, and chalk, other crops can be grown as well.

It's common for livestock farms on these soil types to grow both grass and maize.

Incorporating grassland in a plan with other fodder crops can help to keep the average organic matter content of the soil at a sufficiently high level. Grass/clover is especially suited to crop rotation, since it grows well on the nitrogen-poor soils which often result from growing frequent crops of maize. ▼

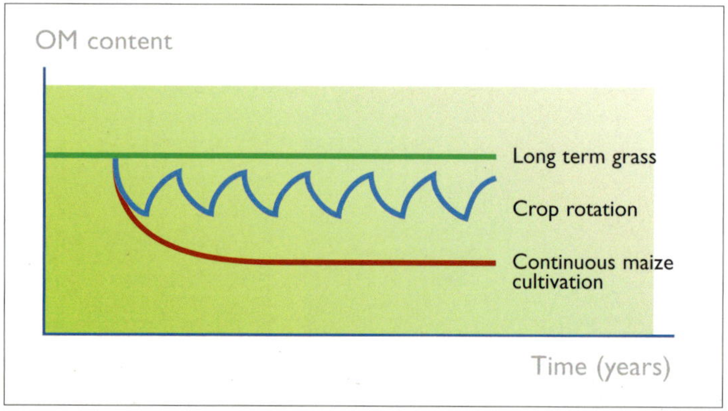

Choices

Make clear choices. If you opt for permanent grass, stick to your guns. Don't take the plough to permanent grassland if it's not needed for grassland improvement. The decision to tear up permanent grassland in order to sow one crop of forage maize will result in huge organic matter losses. There is also a chance of nitrogen being leached. After permanent grassland has been ploughed, the breakdown of organic matter is also much greater than after ploughing temporary grassland since the organic matter content is usually higher.

Crop rotation

Grassland fits well in a crop rotation plan, especially on light sandy soils. Incorporating grass in a rotation with maize and perhaps cereals as well can help maintain the soil fertility. In particular, continuous cultivation of maize leads to a decrease in soil quality, organic matter content and thereby the soil's production capacity. In a crop rotation, there is an increase in organic matter in the 'grassland phase' and breakdown of organic matter in the 'arable phase'.

▲ *Maize usually does well on ploughed-up grassland. The maize crop benefits from the nitrogen freed when the old grass layer decomposes.*

Green manure a must

After growing maize on sandy or chalk soil, you must plant a green manure crop. Grass is also a good idea here. A crop of Italian ryegrass can retain about 30 kg N per ha. You can save this amount of nitrogen when fertilising the follow-up crop after the winter.

An alternative after maize is winter cereals, so that the grass can almost always be sown under better conditions in August of the following year: it gives grass (and clover) a better start, an investment that will be paid back in future years.

Chapter 3 February-March

The pasture awakens

Towards the end of the winter (February-March) the grassland begins to awaken. Determine how well the grass has made it through the winter. Look below as well as above the grass sward. The quality of the soil plays a large part in determining how well the grassland will produce in the coming season. Good management and manuring will give grass an additional boost.

▲ A healthy soil ensures grass kicks off to a good start in spring.

A healthy soil is the basis for healthy grass growth. It is important for the plant to have enough moisture and air in the soil and easily available nutrients. Fine tuning soil quality and mineral management will help to combine good grass growth with appropriate fertiliser application.

Dig a hole

Take a look into the soil by digging a hole. Find a representative area in the field and dig a hole 50 x 50 cm and at least 40 cm deep. It will give you a good idea of the root growth, soil structure and activity of the soil organisms. This information is a useful supplement to the figures from soil chemical analysis.

Soil particles play an important part in an assessment of soil structure. A crumbly soil with rounded crumbs is good. Sharp edges are bad for soil structure. They inhibit root growth, the movement of water and the breakdown of manures. ▼

Assessing root growth

Look at the profile of the wall of the hole. The roots tell a great deal about the structure of the soil. New grassland has roots down to 70 cm. A year later, the vast majority of the roots will be no deeper than 35 to 40 cm. On older grassland, the root depth will often decrease to 25 or 35 cm. There is a strong correlation between the soil management on grassland and root depth. On badly managed grassland with deteriorating soil quality, root depth will decrease more rapidly.

Normally, 60 to 80 percent of the living root mass is in the layers from 0 to 10 cm. The depth of the roots will depend on the grass species and variety; for example, cocksfoot and tall fescue will root deeper than perennial ryegrass and rough stalked meadow grass. This can be a consideration in choosing a grass mixture for drought-prone soil. The roots show whether or not the soil has a good structure. Shallow rooting or kinks in the roots indicate structural problems such as compacted layers.

Dig every year

Make it a habit to dig holes in a few fields every year. It's an effective way to get an idea of the condition and quality of the soil. If you do this regularly on a number of fields, you'll gain experience in examining your soil. It also makes it easier to identify long-term trends.

Types of earthworms

Worms are good indicators of total soil organism activity. They fall into three groups.

Red worm

Red worms are 6 to 15 cm long. They live in the litter layer and break down manure and plant remnants. Red worms prevent the turf layer from matting. They do not like the soil cultivation.

Grey worm

Grey worms are 8 to 14 cm long and greyish, sometimes with some blue or pink. They live in the upper 40 cm of the soil, where they make a criss-cross of channels as they 'eat' their way through the soil. This makes them very beneficial for soil structure improvements.

Common earthworm or lob worm

Common earthworms are 9 to 30 cm long. They make vertical channels and transport organic material downwards. They are capable of breaking through compacted layers. This allows rainwater to drain through faster, and allows air and roots to reach the subsoil better.

Good and bad grasses

- Good grasses: perennial ryegrass, timothy grass and meadow fescue.
- Moderate grasses: rough stalked meadow grass, smooth meadow grass, couch grass, meadow foxtail, yorkshire fog, creeping bent, cocksfoot and tall fescue (pretty good for cutting).
- Bad grasses: annual meadow grass, marsh or bent foxtail, red fescue, sweet vernal grass and velvet bent.

Signal plants

- Watch out for these plants, as they are indicators of less favourable conditions:
- Sheep's sorrel, yorkshire fog > low pH
- Creeping buttercup, marsh or bent foxtail > poor drainage
- Couch grass, reed canary grass > loss of biodiversity

Assessing plant composition

Early spring is the correct time to assess how the grassland has come through the winter. If grassland has suffered over the winter, there can be open swards or lots of bald spots. See if there is at least one healthy plant of perennial ryegrass on average per 10 x 10 cm of surface. If not, or if the distribution is irregular, consider over-seeding or re-seeding.

There may also be enough grass plants, but not of the species you want. Grasses can be divided into good, moderate and bad grasses based on lifespan, sward forming, frost resistance, dry matter production, growth curve, taste, digestibility and nutritional value.

Some winters really take their toll on the grassland. Less hardy grasses will die off. There may be too few healthy tillers of good grasses (especially perennial ryegrass). ▼

If slot seeding is required, do it as early in the spring as possible. Use 30 kg perennial ryegrass per ha for diploid grass or 45 kg for tetraploid grass. Are there lots of weeds in the field? Apply directed herbicide before slot seeding. ▼

Criteria for re-seeding

The decision whether or not to re-seed will depend on the water balance in the soil. It is more difficult to maintain a high proportion of good grasses in wet fields.

Consider re-seeding on wet ground:

- if the percentage of perennial ryegrass and rough stalked meadow grass together is less than 50 percent;
- if there is more than 25 percent couch grass spread across the field or if more than 15 percent is in patches.

Grassland improvement may be on the cards for a field with good water balance if:

- the percentage of perennial ryegrass is less than 50;
- there is more than 20 percent couch grass distributed over the field or if more than 10 percent is in patches.

Couch grass is not desirable in your pasture, since it is not very productive and its palatability and nutritional value decrease rapidly as it gets older. Grassland with couch grass is difficult to manage. It bolts quickly and forms tussocks which smother and overshadow the other grasses.

An increasing proportion of couch grass in the grassland goes hand in hand with a decrease in digestibility of the grass for the cows. ▼

White and purple feet

Perennial ryegrass and rough stalked meadow grass resemble each other in spring. One visible difference is that perennial ryegrass (right) has 'purple feet' whereas the base of the stems of rough stalked meadow grass (left) are plain white. It's easy to see the difference if you compare a handful of each. Another characteristic of perennial ryegrass is that it's very tough just above the ground. You can feel this easily by pulling out a handful of grass. Perennial ryegrass also has obvious leaf ribs, which rough stalked meadow grass does not. The backs of the blades of perennial ryegrass are shinier than those of rough stalked meadow grass.

▲ *Rolling has to be done at the right time. If the ground is too wet, it may be damaged by the tractor and roller. If it's too dry, rolling will have little effect. You have some flexibility in timing by adjusting the weight of the roller.*

Extra attention

Once grass growth is underway, grassland will need some extra attention. Harrowing grassland only makes sense if there has been some matting of the sward. It also spreads cowpats and molehills around. Working the ground with a pasture harrow stimulates new grass tillers. If there is a lot of chickweed growing amongst the grass, a weed harrow is a useful alternative. Harrowing deals a blow to the chickweed, which will then have difficulty becoming established amongst the growing grass. Chickweed will usually disappear after one pass with the harrow.

Rolling the ground makes it firmer when it has become loose, for example after re-seeding. It can also repair the contact between root zone and subsoil, which may be necessary when there has been damage from beetle larvae. Rolling will make for level swards on clay and peaty soils.

A practical method of keeping a dense sward is to spread grass seed with a fertiliser application two to three times a year. Spread 8-10 kg seed per hectare. Cows will trample the seed into the ground, and sparse areas of sward will fill in.

◄ *Field harrows come in many shapes and sizes. The desired effect can be achieved even with the most simple machines. Harrowing is not a way of disguising your failure to tackle moles.*

Working grassland with an aerating machine can be done more often as there are indications that aeration of grassland with compacted soil helps get air into the soil, with positive effects for soil structure and grass growth. ▼

White and red clover

In addition to grass, grassland sometimes contains clover. Clover and grass make a good combination, as both red and white clovers capture nitrogen from the air and fix it in the soil, which benefits grass. This can make a grass-clover sward highly productive where nitrogen fertiliser use is limited.

White clover is vulnerable to cold winters, but the clover's ability to recover is superb. If you only have small amounts of white clover in spring, it's no cause for concern. White clover plants can spread rapidly through the field of grass using their stolons.

Red clover cannot spread with stolons. Thus, the amount of red clover in spring will determine how much is available for the rest of the growing season. Red clover is primarily suited to swards cut for silage, since the plant is not suited to grazing.

▲ A reduction in red clover can be an argument for re-seeding a silage field. Pay attention to the amount of white clover as well: if it's well distributed over the field and comprises about 10 percent of the herbage mass, and if there are still enough good grasses, re-seeding is unnecessary. White clover will take over the red clover's function of providing nitrogen.

◄ Due to its requirement for warmer temperatures, growth of clover starts later than grass. White clover only grows if the temperature is above 6 - 8°C. In spring it may look like the white clover in the grassland has come off worst, but it usually improves after the first cut. If you're really concerned about the amount of white clover, consider taking an earlier, light first cut.

Weeds and herbs

Besides good grasses, moderate grasses and clover, other plants inhabit the sward. Excellent for biodiversity, but if they inhibit pasture yields or reduce the nutritive value or acceptability of the forage, they are usually labelled as weeds. There are also weeds that are toxic or cause health problems in livestock. With good grassland care, you can usually prevent weeds becoming a problem. If weeds are rearing their heads it can be a sign than conditions aren't optimum. The presence of buttercups, for example, indicates the land is too wet. When you assess the land in spring, look at the weeds that are growing and decide whether killing them is necessary, based on species and density. Chemical sprays are the most common ways to kill weeds. This can be by whole-field or spot application, depending on how the weeds are distributed. Spot-spraying saves money and is better for the environment. The effect of sprays on many weeds, e.g. docks, is stronger in autumn than spring. Still it may be necessary to intervene if the grassland comes out of winter with too many weeds.

The red deadnettle is one flowering in the early-spring. It often grows into a field from headlands. ▼

Chickweed takes its opportunity in open swards or bald patches. This weed can grow at temperatures below 8°C, when grass growth is stagnant. Since chickweed grows quickly it can smother the grass. A treatment with the weed harrow in spring can put chickweed at a disadvantage. It can be completely killed off with spraying. ▼

Docks are best dealt with by digging up the entire plant, including the long tap root. Organic farms in particular keep grasslands free of docks this way. They can be weeded all year round. ▼

Dwarf nettle. Nettles like to grow in nitrogen-rich spots.

Shepherd's purse. A very common plant found in most of the world.

Knotweed. In the sixteenth century it was commonly fed to pigs that had lost their appetites.

Celandine. Celandine blooms very early in spring. It especially likes canal banks and other moist places.

Take care with new grass

Dealing with weeds deserves extra attention on newly sown grassland. Cultivating soil before sowing gives weed seeds a chance to germinate. As fields sown the previous autumn emerge from the winter, chickweed is usually the problem weed. When sowing in spring, shepherd's purse and fat hen can be a problem. Chickweed smothers young grass and shepherd's purse displaces it by forming rosettes. Both weeds cause open swards and production loss. Prevent bare patches in the field and spray if necessary for chickweed and shepherd's purse as early as possible, starting from 1 March and a few weeks after sowing.

Spray at the right time

Sometimes spraying is necessary in spring. Annual summer weeds get the best opportunities in a spring-sown sward. The product choice depends on the weeds. Shepherd's purse and chickweed are easy to kill in early spring. Do wait for good growing weather to spray. For other weeds such as dandelions and buttercups, it's better to wait until April to spray.

A disadvantage of slurry is that it contains a lot of water in relation to nutrients and is thus relatively heavy. This means a greater risk of damage to the soil from machinery if the land is still too wet. You can limit soil compaction by using an umbilical hose. An advantage of slurry is that it is a 'complete' type of fertiliser rich in nitrogen. Thus it makes a substantial contribution to satisfying the grassland's need for fertiliser.

Starting with manuring

There are two types of manure for grassland: animal manure and mineral or artificial fertiliser. One advantage of using manure is that it not only applies mineral and trace elements to the soil but organic matter as well. Farm yard manure provides more organic matter than slurry. Another characteristic of manure is that not all the nutrients it contains are directly available to the plant, especially so with nitrogen. The temperature of the soil and moisture conditions influence how much organically bound nitrogen is released.

Fertilising with manure is legally allowed after 15 January. It's a good idea to spread manure as soon as possible after that, since organically bound nutrients are released gradually. The spreading time will be determined by the soil's carrying capacity. Spreading in overly wet conditions can compact the soil, and excess rainwater can leach away minerals into groundwater. Pastures with a later cutting date should be fertilised later (or even not at all before the first cut); this will limit the risk of a heavy first cut with a low nutritive value.

The tracks on this grassland are from spreading slurry while the carrying capacity of the sward was insufficient. Getting on the land too early will damage the sward and soil structure. ▼

Improving utilisation

Slurry applied to grassland in spring, releases on average 50 to 60 percent of total nitrogen as available nitrogen to plants in a few weeks.

In practice, the spring application of cattle slurry varies from 20 to 30 cubic metres per hectare. Due to the gradual release of nutrients and the lower risk of leaching, it's a good idea to get a large portion of the available slurry on the grassland in early spring. A smart way to limit the nitrogen loss on sandy soil, which is vulnerable to leaching, is to spread the spring application over two rounds, for example 15 cubic metres per hectare each time.

The size and pressure of your tyres will play a large part in the pressure a machine exerts on the soil. Ideally, work with low-pressure tyres, with a pressure under 1.0 bar. Low-pressure tyres can be recognised by their 'chubby cheeks'. Not all tyres are made for such low pressure. ▼

Puzzler

What can you see here?

This is a case of improper slurry injection. The injector made too shallow grooves and the application rate was too high. Result: the slurry flowed in channels along the ground and smothered the grass plants, creating an ideal situation for couch grass and weeds to spread. In The Netherlands, spreading slurry this way also does not usually meet legal requirements, since it fails to reduce the ammonia emissions.

▲ *Liquid fertiliser can be very evenly distributed.*

Puzzler

What went wrong here?

The grass in the photo was given slurry and artificial fertiliser perfectly on time. Yet it is yellowish-brown and is barely growing. The reason: the weather was dry and cold for weeks on end. The grass was insufficiently able to take up the nutrients and grow properly. In this case, application of additional nitrogen isn't necessary. Once the weather turns, the crop will recover. Thus, applying fertiliser at the right time is no guarantee for 'perfect' grass growth. The weather conditions after application have a role to play too.

Artificial fertiliser

Applying artificial fertiliser to grass-land is usually done to supply nitrogen, but it is often used to supply sulphur and phosphate as well. When animal manure cannot be applied, for example on new grassland with a low carrying capacity, it may be necessary to supply the required nutrients in the form of fertiliser.

The right time to apply chemical fertiliser is an important decision, especially with regard to nitrogen. If you apply too early, you risk loss of nitrate due to leaching. If you apply too late, you will suffer unnecessary yield loss at the first silage cut. The ideal stage at which to apply fertilizer is when the sward becomes active after the winter, which is mainly determined by soil temperature.

One way to calculate this is the T-sum (temperature sum). The T-sum is the accumulated mean daily temperatures above 0°C after 1 January. Applying nitrogen between T-sum 200 and 300 will yield the highest grassland production on average. You can find the T-sum calculator for English areas at www.farmwest.com (climate tab), and the Netherlands (in Dutch only) at www.nmi-agro.nl.

Artificial fertiliser usually supplements the value of applied manure. A clever choice of nutrients used in a blend (mixture of fertiliser types) can help limit the number of trips with a fertiliser spreader. ▼

Start signals

An even better way than working with the T-sum is basing the application time on signals directly connected to the biological activity in the soil. One good method is to take a look at the soil. Visible white root tips are a sign of spring growth. You can also wait until you see other plants in or near the grassland starting to flower, such as willow (catkins) or coltsfoot. Find out which plants you can use as signals in your own situation for the right time to apply fertiliser.

Dividing fields

The amount of nitrogen you have to spread in the form of artificial fertiliser before the first silage cut depends on the slurry applied, the soil's mineral nitrogen-supplying ability, the planned crop use and the planned nitrogen level per hectare per year. In practice, it is difficult to calculate and spread a different application of nitrogen for each field. Find a practical middle ground, and divide the parcels into two or three groups, giving each a different application. You can use RB 209 (English) or www.bemestingsadvies.nl (Dutch only) to determine how much nitrogen you can apply to the first cut in relation to the desired total nitrogen application per year.

▲ *Taking the soil's temperature is a reliable method for determining the time to apply nitrogen. Grass starts to grow once the soil temperature has reached 8°C. The T-sum describes temperatures above ground and is also an indicator for soil temperature. However, measuring soil temperature directly gives more certainty.*

Willow catkins are a sign that the grass has also initiated growth. Apply nitrogen once you can drive a machine onto the land. ▼

Coltsfoot (below) starts to flower when the T-sum is 180. With very mild winters and after harsh winters the flowering phase of coltsfoot is a better predictor of the right application time than the T-sum. Flowering celandine is also a good indicator. ▼

▲ *In spring it takes at least fourteen days before grass plants sprayed with glyphosate die off completely. Wait before you cultivate, otherwise the odds are high that couch grass will re-emerge.*

Grassland improvement

Current NVZ regulations, where over 80% grassland derogation and 250kg/N/ha is applied for, only permit the re-seeding of grass on sandy soil in spring. After 10 May, the sward may not be destroyed either chemically or mechanically. On clay and peaty soils the sward may be destroyed in spring, summer and autumn.

If grassland needs to be improved, you can choose between sowing grass seed into a desiccated (dead) sward without cultivation, or in a seedbed after cultivation. Re-seeding combined with ploughing or digging is often tricky, especially on heavy clay and peaty soils. Heavy clay makes it difficult to prepare a good seedbed, and peaty soil usually has insufficient carrying capacity. In these situations, it's preferable to over-seed into a desiccated sward. The land must be level, however, and there must be no further need for levelling or drainage. Re-seeding on tillable soil is a good option if the soil structure, drainage and soil levels are in order. Whether considering over-seeding or re-seeding in combination with cultivation, the old sward should be killed off first, especially if couch grass is present.

Spraying with glyphosate is the best way to kill off a sward. Spray preferably in the spring. Couch grass must be growing and have a leaf length of at least 15 centimetres. A minimum average daytime temperature of 8°C and a minimum average night temperature of 5°C are also required. As with other chemicals, spray carefully following the instructions. ▶

Ploughing, digging and rotary cultivation

Cultivating the ground begins with breaking up the old sward, with a rotavator, discs or rotary harrow, to a maximum depth of 5 cm. Next the ground is cultivated deeper, either by ploughing or rotary digging. Rotary digging has the advantage that more nutrients and organic matter remain in the top layer of soil than when ploughing. Another advantage over ploughing is that there is no risk of leaving a plough pan, which prevents deeper penetration by roots.

An advantage of ploughing is that the soil is turned over, so weed seeds are covered deep down. This limits competition from weeds in the new sward. Limit ploughing depth to 20 or 25 cm. If you use a furrow press at the same time, you'll have a consolidated seedbed. After ploughing or digging, the ground can be levelled if necessary. This is best done using laser equipment.

After cultivations, necessary adjustments can be made to bring the nutrient status of the soil to the desired level. Often liming is required. Various liming products are available.

▲ *A rotavator cuts the old turf layer into small bits. Make sure they are no larger than 5 cm. Larger pieces are more difficult to plough under neatly and won't break down as quickly.*

An eco-plough works at a depth of 12 to 18 cm, less than a conventional plough. This prevents nutrients and organic matter from disappearing too far into the lower soil depth. The disadvantage of working with an eco-plough is that it won't loosen deeper compacted layers. ▼

Low load bearing soils take better to digging than ploughing. Even if the aim is to mix different layers of soil, digging is preferable. A digging machine will mix the soil to a depth of 10 to 30 cm. ▶

▲ *Combining seedbed preparations with sowing is effective, for example by working with a rotor-head harrow and drill. Beware of too much and too intensive cultivation: a seedbed can be too fine.*

Seeding

With slot seeding, grass is sown into an existing sward. If the grass cover has been sprayed to desiccate the old grass, you're re-seeding. An advantage of slot seeding is that less organic matter is lost. slot seeding is best done in moist soils. If you intend to inject slurry in the ground, do it one to two days before slot seeding. Applying earlier can cause problems by capping the soil.

Grass that germinates in the slots made by the slot seeding machine is highly sensitive to drought. It's desirable to keep the soil moist after slot seeding. Harrow the field after slot seeding, and irrigate if necessary.

Seedbed

When re-seeding in combination with cultivation, time everything as closely together as possible so the soil does not dry out. The seedbed must consist of newly compressed subsoil with 1.5 to 2.5 cm of loose, crumbly soil on top of it. When sowing 1.5 to 2.5 cm deep the grass seed will lie on the pressed-down layer.

Broadcast sowing of grass seed (not in rows) will most quickly result in a dense sward and will give fewer problems with weeds.

Over-seeding is an inexpensive form of grassland improvement, but it has its risks. The chance that the seed will not 'take' is greater due to the lack of a real seedbed. Moisture levels can sometimes be inadequate. The chance of weed ingress is also greater. ▼

Broadcast coulters on a seed drill distribute the grass seed evenly and not in rows, compared to a drill with conventional coulters (see photo). ▼

Grass-clover

If you've chosen to sow a grass-clover mix, the seedbed will need some extra preparation. A fine, firm seedbed is more important for clover seed than for grass, as is the requirement that clover should be sown at the correct depth. The ideal sowing depth for clover is less than for grass, 1 to 1.5 cm. It's better to use a drill that can simultaneously sow clover separately from grass seed.

Clover is more susceptible to drought than grass immediately after sowing. The seed can absorb dew and swell, then dry out during the day, which destroys the emerging seedling.

Grass and cereal

When sowing grass-clover in spring, you can also sow 25 to 50 kg of cereal (spring barley or oats) at the same time as a cover crop. Cereal helps you achieve a fast, closed crop that gives weeds very little opportunity. Cereal roots also have a positive effect on the soil structure, and encourage soil organisms. After cutting, the cereal disappears from the grass-clover mix.

Grass and peas

Combining grass with (semi-leafless) fodder peas will add protein to the forage. Peas take well, and the grass will take over after the first silage cut.

▲ *Combine grass with both white and red clover when you solely intend to cut a field for silage.*

To increase the chances of a good clover-grass crop, consider a cereal as an intercrop. After the cereal is cut the soil will be low in nitrogen, which will give the clover a good start. Cereal crops also improve the soil structure. You can sow grass-clover after a cereal harvest until 1 September. ▼

◀ *Rolling seeded grassland with a Cambridge (ridge) roller can be worthwhile, especially if it's dry around seeding time and if the seedbed is somewhat coarse. A Cambridge roller will break up clods and consolidate the ground but without the risk of dust blowing away associated with using a smooth roller.*

Chapter 4 April-May

The time of grazing and taking the first cut is one of the peak times of stress in grassland management. There are many possible choices, ranging from starting early with cutting, to cutting that serves the grazing to mowing the first cut entirely.

▲ *In a normal spring the first cut will provide a significant part of the forage supply, in both quantity and quality, for the next time the animals are housed.*

Just as every spring is different, so is every spring's grass growth. Temperatures, light, moisture and nutrients all vary. Try to get an idea of how the grassland is developing by walking all your fields weekly once the grass has started to grow. When you've had a few night frosts, the grass can have a purplish sheen: the cold reduces the availability of phosphate for the grass plant, which results in the purple colour. Cool but sunny weather, usually with cold nights, can raise the sugar levels in the grass. You can feel the increase in sugar by crushing a handful of grass shoots between your hands at the end of the afternoon – the sugar will make it feel sticky.

A tour around the fields once a week will give you a good idea of how the grassland is coming along. ▼

Signal areas

Irregularities in fertiliser application show up when the grass grows. Stripes in the grass can indicate uneven distribution of the slurry or fertiliser. If you've used manure there will be some overlapping, and thus higher manure dosage, especially on tapered fields and some of the headlands. The grass will have a head start on the more heavily fertilised areas of the field. Use this knowledge and keep an eye on the more heavily fertilised areas. When that grass lodges, it's a sign that this will happen soon to the rest of the field, and that it's time to cut.

Dandelions contain many minerals such as calcium and iron. Since they have a reasonable value as fodder you don't need to go to the ends of the earth to eradicate them. They can multiply over the years, however. There will be a loss in yield if their proportion in the dry matter rises to over 25 percent. Dandelions like well fertilised grassland. They can be easily attacked by a chemical spray. Carry out in early April, before they bloom. ▼

Look at that sod

Assess spring growth both above and below the soil surface. Lifting a 20 x 20 cm sod can provide a lot of information. Count the roots you see on the underside of the sod. Twenty roots are not many, and two hundred is a lot. Cut through it at 10 cm down and repeat the count. Sixty roots is low, four hundred is high. Too few roots can indicate compacted soil. Look at the shape of the roots as well. Are they straight, and can they grow uninterrupted, or do they point to structural problems? Look on the underside for worm holes and worms themselves. Three holes are not many, and fifteen is a lot. A compacted soil can recover if there are a lot of worms and wormholes.

▲ *All sorts of aids are available for measuring grass height. Many livestock farmers get on just fine with a scale on their boots.*

Estimating yields

In order to manage grassland well, you have to know how much grass is present. This is the basis for decisions such as when to turn out the livestock onto the land, or when it's time to mow. However, determining the actual yield isn't that easy. As long as the grass is in the field you'll have to make do with an estimate. The dry matter yield will depend on the height of the grass, density of the sward and dry-matter content, among other things. Measuring the height is relatively easy. There are special grass height meters available, but a measuring tape or ruler will do just fine.

Determining the density of the sward is trickier. It can only be done by eye and takes some practice. Italian ryegrass and tetraploid grasses usually have a less dense sward than diploid perennial ryegrass.

A grass height meter gives a good idea of the grass height. A tool like this can be of great help to give you a feel for what's on the ground. ▶

A silage field has a more open appearance than a field which has been regularly and intensively grazed. Tetraploid perennial ryegrasses reinforce this picture. ▼

Watered-down yield

The dry-matter content of cut grass will depend on several factors including growing conditions up to the time of cutting, grass species, other plant species present in the meadow and the season.
The dry-matter content can vary from 10 to 25 percent. Thus there can be huge differences in dry-matter yield between grasses of the same height. In a 'normal' crop the dry-matter content will often be between 15 and 20 percent. The dry-matter content of perennial ryegrass is on average 16 percent when cut, and tetraploid varieties have lower dry-matter content than diploid. This figure will be several points lower for water-loving species such as rough stalked meadow grass. White and red clover also have lower dry-matter contents than perennial ryegrass.
Incorporate weather conditions and humidity of the soil in your estimate. The same grass on wet ground in cloudy weather has lower dry-matter content than in drying conditions. Fertiliser also plays a part in your estimate. The higher the level of fertilizer, the earlier you can take a heavy cut off the land.
Experience, however, shows that estimates of heavily fertilised grass are often above the actual yield. The reason is a decreasing dry-matter percentage with high fertiliser levels.

The dry-matter content of the different grasses varies widely. The weather and the season also have an influence on how 'watery' the grass is. ▼

◀ *A field of grass with a lot of rough stalked meadow grass often has a dry-matter content of no more than 12 percent. This grass contains much more water than perennial ryegrass. This means that an impressive looking swath of grass may have little left over after cutting and wilting.*

Guide on estimates of herbage mass for grazing and herbage yield at cutting

(kg dm/ha)

Grass height (cm)	*Herbage mass*	*Herbage yield on cutting at 5 cm stubble*
9	2175	825
10	2285	935
11	2400	1050
12	2520	1170
13	2645	1295
14	2775	1425
15	2910	1560
16	3050	1700
17	3195	1845
18	3345	1995
19	3500	2150
20	3660	2310
21	3825	2475
22	3995	2645
23	4170	2820
24	4350	3000
25	4535	3185
26	4725	3375

Correction factors:

- *More than normal grass density: estimate + 10%*
- *If 25% are herbs or water-loving grasses: estimate – 10%*
- *Tetraploid grasses: estimate – 10%*
- *First two cuts after re-seeding: estimate – 10%*

▲ *Avoid puddles on newly sown grassland. Take care of problem areas by digging a small channel into a drainage ditch or stream so rainwater can drain away.*

Take good care of new grassland

Take good care of newly sown grassland. Keep an eye out for the new grass and any clover receiving competition from weeds or other grasses. Some weeds such as goosefoot and nightshade can be tackled by topping them with the mower. They will lose their competitive edge over grass, and will usually die off quickly. Cutting with an ordinary mower is better than with a topper since the latter often only cuts off the leaf tips, and can leave too many tracks. Other weeds, such as chickweed and shepherd's purse, may require spraying. Chickweed races ahead if the ground is capped.

Treat new grassland with a roller if necessary to improve firmness. This can be done with a Cambridge roller before the grass comes up, and after germination with a smooth roller.

A light cut or and early start to grazing will help the young grass to form new tillers, as long as it is not poached. Keep the stocking rate low. If you have sheep, let them do the first grazing.

To prevent damage to the sward and compaction of the soil, it's better not to use slurry on new grassland before the first two cuts.

Meadow grass sometimes competes with newly sown grass. With good grassland management it won't have an opportunity to dominate the new grass. This means killing weeds, timely grazing, fertilizing sufficiently and not taking heavy cuts. ▼

Rolling new grassland can help firm the soil. A very loose soil will slow down tiller formation, giving weeds more opportunities. ▼

Meadow birds

Meadow birds are inextricably connected with grassland. Many livestock farmers are fond of them and take measures to ensure the birds can hatch out their eggs and bring their young to adulthood. They sometimes go to great lengths to help out, especially for rare species such as the black-tailed godwit.

Meadow birds lay their eggs from mid-March through April. To prevent the nests being destroyed by your work activities, they have to be located and marked. Often volunteers can be found to do this. But finding and marking the nests alone is not enough. The chicks are very vulnerable for the first few days after hatching and are in considerable danger of being killed during mowing. If they stay out of the mower's way, their survival still isn't assured, because in a habitat of only cut grassland they are an easy prey for all kinds of predators.

Mowing different sections at different times or leaving wide swaths of grass uncut are ways to give young meadow birds a start until they can fly.

Place a stick 5 meters in front of and 5 meters behind a nest. This will prevent clever predators like foxes and crows from detecting and robbing nests with the aid of a nest marker. ▶

By leaving wide strips of grass (at least 10 meters wide) uncut, you'll give meadow bird chicks better chances of survival. If you leave these strips uncut for at least two weeks, there will be enough grass on the rest of the field to offer protection to the young birds. ▼

Curlew eggs. Like the black-tailed godwit, redshank and others, curlews are endangered meadow bird species in the Netherlands. In meadow-bird habitats, there are often subsidy schemes to compensate farmers for the costs of helping out the birds. ▼

Grazing systems

The most inexpensive method of utilising grass with your cows and sheep as high-quality fodder is to let them retrieve it themselves in the meadow. A grazing system will ensure that they always have fresh grass available.

Roughly speaking, there are two grazing systems, continuous and rotational stocking. In continuous stocking the cows are on the same field for a longer period of time. With rotational stocking they go to a new piece of pasture more frequently. You can compensate for the fact that grass grows at different rates at different times of the year by varying the area to be mown.

In rotational stocking the frequency of moving to a new pasture (or part of one) can vary from once every six days to twice a day. Which system will best fit the farm will depend on, among other things, the farmer's personal preference. How the land is divided into paddocks, and whether a lot of land is near the parlour or not, can also play a part.

Grazing younger animals behind milking cows is a practical way to use grassland efficiently. The milking cows always have tasty and highly nutritious grass available. Yearlings can manage with lower-quality fodder, and can eat what the cows do not. One disadvantage is that the young animals must change fields often. ▼

Strip grazing is a form of rotational grazing in which the cows receive a new strip of grass each day (or twice a day) with enough on it for just that grazing period. This minimizes hoof damage of the sward, which allows the grass to grow back faster. ▼

When should I turn out?

Following winter, introduce cows to fresh grass gradually. The rumen bacterial population will have more time to adjust to the change in diet. Start grazing as soon as there is about 2400 kg dry matter present per hectare, this will provide approximately 900 kg of dry matter per hectare for the animal. Turn the cows out for just a few hours a day during the first few days. They'll keep eating the majority of their ration in the shed. Increase the hours spent grazing to the desired number over following two weeks.

With rotational stocking, your goal is to have the cows coming into new paddocks once it has 3200 kg dm/ha herbage mass. Especially in the first few months of the growing season, before the first cut, this is often tricky to achieve. Grass grows very quickly in the spring, so you will soon have more grass than the ideal grazing yield.

Grazing gridlock

Livestock farmers practising rotational stocking sometimes get stuck in early May. The grass on ungrazed paddocks has grown too far to allow cows onto it. Regrowth is not yet available. You'll have to graze the cows on the first grazed paddocks. But if you've started grazing too late, there won't be enough grass on these areas. It's important to bear this in mind when you start grazing. Try to have as large an area for grazing available as possible after the first cut, so that you can go back to the first grazed paddock after about 2½ weeks. After that, the aftermaths will be ready for grazing. Some dairy farmers choose to mow the complete first cut to prevent grazing problems in spring.

There are various stocking systems possible, and they can be used one after another. For example, it can be convenient to continuously stock before the first silage cut. It's easier to organise. When the silage cut has been taken off the land, you can go to rotational stocking. ▼

Pros and cons of grazing systems

	Grass production	Loss	Taste	N utilisation	Labour
Rotational (traditional)	+	0	0	0	+
Continuous, set stocking	-	-	0	0	++
Rationed grazing	+	+	+	0	0
Strip or paddock grazing	++	++	++	+	0
Siesta grazing	+	+	0	++	-

- = less favourable; 0 = neutral; + = positive; ++ = very positive

▲ *In rationed stocking, the cows get a new area of the paddock added every day, so they always have fresh grass. It can also be an emergency measure if you want to turn the cows into a paddock with too tall grass. This will limit losses from grazing. Compared with strip grazing, rationed grazing increases the area available to the cows every day. You still have the disadvantage of rotational stocking, a greater delay in re-growth.*

Rotational stocking

In rotational stocking, the paddock area is adjusted to the size of the dairy herd and grass intake per animal per day. Rotational stocking is a flexible system in which you can skip a paddock if the grass growth is more than anticipated. You can keep turning the animals in on grass of the right length and good nutritional value. A disadvantage of this kind of correction is that you leave a paddock for mowing that was fertilised for grazing. You may lose yield and nutritional value. You can retain nutritional value by taking a somewhat lighter cut.

When working with rotational stocking, the main thing is to balance grazing and cutting. Cutting smaller areas is one method to ensure that more and more paddocks with the right grazing height become available.

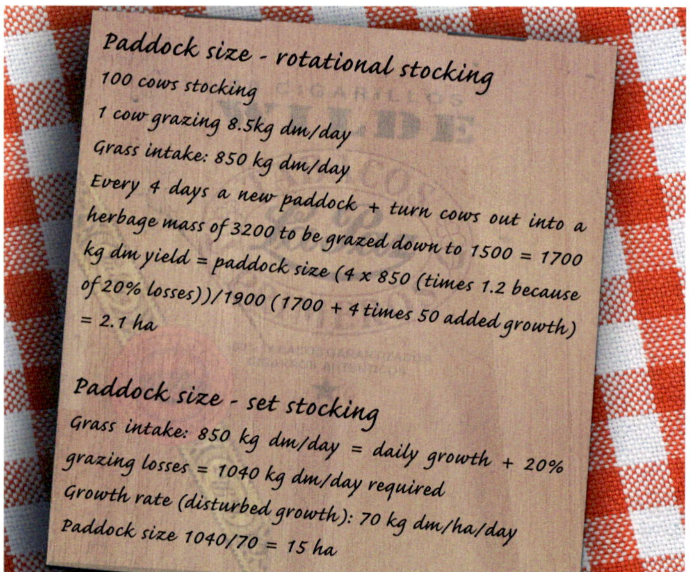

▼ *Rotational stocking means more fences on the land. If you work with temporary fences, you can remove them quickly for mowing or fertilising if you need to.*

Modern set stocking

In modern-day set stocking, a livestock farmer divides the grassland into blocks, on which the cows graze for several weeks at a time. After two to six weeks the cows move to a new block, cut one to two weeks before. This means the cows are always in a pasture with grass 8-10 cm long.

Varying the supplementary feed when the animals are housed is a tool for adjusting the grass intake from the field and thereby the grass height. When the grass in set stocked paddocks becomes too tall the losses increase considerably.

On the other hand, if the grass is grazed too short, the growth rate and nutritional value will suffer.

Set stocking is less labour-intensive for the farmer. Among other factors, the land needs less frequent cutting and fewer applications of chemical fertiliser. In modern set stocking, grass intake and milk production are comparable with those achieved in rotational stocking. The total grass production is 10 to 15 percent less than with rotational stocking, however. For extensive livestock farms with a lot of land, this need not be a problem.

▲ *Cows are often calmer under continuous stocking. Under rotational stocking, the animals frequently anticipate movement to a new pasture. Continuous stocking is also easier for the farmer, since there are fewer decisions required.*

◄ *With set stocking, you need to keep an eye on how fast the grass is growing and be able to vary the amount of supplementary feed or the grazing area whenever necessary. The aim is to balance daily intake of grass with daily grass growth.*

▲ *Making the first silage cut at a higher yield than 3500 kg dry matter per hectare does not have to result in a drop in digestibility. There is a chance, however, that the crude protein content will decrease. Although it goes with an increase in crude fibre content, this can give it more structure.*

The first silage cut

The first silage cut usually makes a significant contribution to the winter forage supply in both quantity and quality. With an eye on availability of grazing cuts, it is practical not to mow the entire first silage cut at once. Mow a few paddocks early once there is 2500 to 3000 kg dry matter yield. This will provide a regrowth that cows can graze within a few weeks. If you don't have to or want to take grazing into consideration, cut the entire first cut at an optimum dry matter yield of 3000 to 4000 kg.

Grass grows very quickly in spring and it's not always possible to cut at the right time. It's usually not a problem for the first cut when the grass keeps growing a bit longer. In April and May, ryegrass shows few signs of ageing so long as it doesn't fully start to flower. Its nutritional value will stay the same. Thus, as far as nutrition is concerned, it's not a problem to let the first cut grow to a dry-matter yield of over 4000 kg.

If the grassland consists largely of mediocre grasses you will be less flexible with the time of mowing. The nutritional value of these grasses deteriorates faster with age. Two weeks of wet weather in mid-May can have enormous consequences for the nutrient content of the early-flowering grasses. Mow when there is 3000 to 3500 kg dry matter yield in the field. ▶

The correct time to cut

The weather conditions around the time of mowing are very important for retaining as many nutrients as possible in the ensiled grass. Mowing during sunny weather with a light breeze is ideal. The grass can be brought in from the field with a dry-matter content of about 40 percent a day later. Leafy grass with this dry-matter percentage is easy to process and consolidate in the silo. The conservation process will go smoothly, and there is less chance of secondary aerobic spoilage.

Of course, you can't always mow in ideal weather. Do try to achieve dry matters above 35 percent with a short field period of two days at the most. Promote drying by working with a mower conditioner, which speeds evaporation by removing some of the wax layer on the grass plants. A wide-spreader on the mower will ensure that the grass is well distributed on the field and will dry evenly.

Grass is sticky in the afternoon

Mown grass with a high sugar content sticks to your fingers when you grip it. You'll achieve the highest sugar content in the grass by cutting in the afternoon. High sugar content will help silage fermentation. Grass plants use the energy of sunlight to make sugars, which are used by the grass at night. This is why sugar content is lowest in the early morning and highest at the end of the day. The sugar content of ready-to-mow grass varies from 50 to 200 grams per kg dry matter. Sunny weather and cold nights will give you the highest sugar levels: the grass makes a lot of sugar during the day, and uses very little at night. If you're trying for as high a sugar content as possible, don't wait too long to cut, and don't cut in the evening. If grass isn't sufficiently wilted (i.e. dead) by nightfall, it will continue to utilise its sugars.

◄ *With a well-adjusted tedder, you will prevent the grass mixing with soil. Tedder tines must stay a few centimetres above the ground so that they can leave it undisturbed but lift up all the grass.*

▲ *In grass-clover, the clover can supply the complete nitrogen requirement after the first cut and you won't need to apply any nitrogen. If you do, you'll be making less use of the clover's nitrogen-fixing ability and the proportion of clover may even decrease.*

Manure according to plan

After the first grazing and silage cuts have passed, you have an opportunity to replenish the sward with manures. When a silage cut has been taken for which fertiliser was previously applied, then determining the amount of chemical fertiliser and/or slurry needed is easier. You can fertilise following the plan you made in the spring. When spreading fertiliser nitrogen, bear in mind the mineral nitrogen in the manure you've applied, or have yet to apply (see table).

Often you can't harvest what you fertilised for. For example, if grass is growing quickly, grazing areas will need to be left in order to be harvested later as silage cuts. Especially with regard to nitrogen it's wise to correct this at the next cut. This can be done by a maximum of 25 percent. For example, if you've applied 25 kg N per ha too much on the first cut, lower the N-dose of the second cut by 6 kg per ha.

◄ *A properly calibrated fertiliser spreader is the first requirement for properly applying fertiliser. Spreading fertiliser accurately on grassland is always tricky. Modern aids based on GPS technology can help a lot to improve the distribution of manure and fertiliser.*

Active kg N per m^3 from slurry applied on grassland using a grassland injector after the first cut

Cut after application	1	2	3	4
Cattle slurry	1.1	0.7	0.3	0.2
Pig slurry	2.0	1.2	0.4	0.3
Sow slurry	1.2	0.7	0.3	0.2

Slurry injection for the first cut: effect of 1st cut + 10% and 2nd cut – 10%.
Drag hose spreader: effect in 1st cut similar to injector on 1st cut.
On an annual basis the effect is – 10% (due to surface application).

Manure

Since the nitrogen from manure is released gradually, apply the majority of the available slurry in the first part of the growing season. Applying after the first cut silage is wise. But watch out: grass stubble re-growth after mowing is sensitive to heavy machinery. Wait a few days after mowing before applying slurry, to prevent slowing down the re-growth as much as possible. Grass plants are at their most vulnerable when they have not quite emerged; you don't want to crush the growing points. If leaving a tall stubble, it's generally not a problem to fertilise shortly after mowing. Be careful in dry, overcast weather. Waiting longer than a week after ensiling, however, will contaminate the grass that is growing back.

▲ *When using fermented manure (digestate), bear in mind that the nitrogen is released more quickly in this form. This is because the proportion of mineral nitrogen to organically bound nitrogen is higher. The amount of the variation will depend on the fermenter and the co-fermentation products used.*

A silage cut of 3500 kg dry matter per hectare removes about 100 kg of potash from the grassland, which must be returned. On average, a tonne of slurry contains 7 kg potash. An application of 15 cubic metres of slurry per hectare after the first cut is thus usually enough to meet the potash requirement. ▶

Salt as palatability enhancer

Although grass does not need sodium to grow well, sodium salt does ensure the grass will taste better. Research has shown that fertilising with sodium results in a longer grazing period and higher grass intake by the cows. A beneficial side effect is that sodium inhibits potash uptake, giving you less of a chance that your grass will contain too much of it. Cows eating this grass will also be satisfying their sodium requirements.

Agricultural salt is usually the easiest and cheapest form of sodium to spread. But there are also combined fertilisers on the market with which you can combine the sodium requirement with a nitrogen fertiliser. This is also true for trace elements such as selenium.

After the first cut

During June and July, many grasses have a strong natural tendency to flower, making it very tricky to determine the cutting time for the second and third cuts. If you wish, you can get silage with a better structure in this period. It's also the time to influence the amount of clover in the sward. Grass in Environmental Reserves requires its own management.

▲ *'Heading' often determines when the second and third cut is best taken. To improve structure in the silage, cut in June-July.*

Growth and flowering

After the first cut, conditions are often favourable for a faster re-growth of the grass. The growth rate of grass during May and June is generally twice that of July-August. Starting in May many grass species have the tendency to flower, influenced by day length and temperature.

Allowing the grass to set seed is undesirable since it's at the expense of nutritional value. But it does have an advantage: the flowering of grasses and other herbage offers new opportunities to assess the grassland. Flowering grasses and other plants are often more easily recognised.

Rough stalked meadow grass grows well in the spring, but has poor re-growth after the first cut. Fields with a lot of rough stalked meadow grass can be recognised after the first cut by withered, lifeless stubble. After a few weeks, the rough stalked meadow grass can be recognised as a very fine, often dark green bottom layer in the sward. ▼

Docks

One of the problems with docks is the longevity of its oil-rich seed. It can remain in the soil, capable of germinating, for sixty years. This means that it often raises its head during grassland improvements. To prevent docks spreading further over a farm, it's important to prevent plants from flowering. Mow or clear a field as soon as this threatens to happen. Docks have a taproot that can reach 1.5 meters in length. In their second year, they branch out, creating a dense group of plants after a few years. Full-grown dock plants have a life span of over five years.

Docks are undesirable in grassland because they smother grass and lower its nutritional value. They are also not liked by cows, especially the older plants. Grazing animals will avoid these plants for as long as they can. In meadows with a lot of clover, young docks can sometimes play a useful role, in that the tannins they contain help prevent bloat.

▲ *Docks in verges or the banks of ditches along headlands often have the opportunity to flower, thus 'contaminating' the grassland with dock seed.*

◀ *Grazing influences the botanical composition of grassland, since, among other things, manure and urine are unevenly distributed over the land.*

The underside of red clover leaves are hairy giving it a velvety texture. White clover's leaves are hairless, making it easy to tell white and red apart even if it's not in flower. ▶

Grass species in the flowering phase

Perennial ryegrass

Perennial ryegrass in bloom is recognisable as an ear with eight to ten tiny flowers, and the zigzag pattern of the ear axis.

Italian ryegrass

Italian ryegrass resembles Perennial ryegrass. The ears of flowering Italian ryegrass have needles. During mild winters, Italian ryegrass will sometimes keep growing, but it is not terribly hardy. It has limited endurance: after a few years it will result in an open sward.

Timothy grass

The cylindrical ear tuft with its fork-like spikelets makes timothy grass in bloom easy to recognise. Young timothy grass is very tasty to cows, but not once it blooms. It is very winter-hardy, but does not form a good sward.

Meadow fescue

Meadow fescue somewhat resembles tall fescue (far right), but the two can be easily distinguished when flowering. Meadow fescue likes soil that is not too dry.

Cocksfoot

Cocksfoot is a very coarse-leaved species, and is easily recognised by its flat stem with clear ligules. Cocksfoot grass also has a coarse flower plume. This species can withstand drought and mowing regimes outstandingly well. It also does well in shade and is therefore often found under trees. Cows are not keen on cocksfoot grass, especially when it is older, as its acceptability declines.

Smooth meadow grass

Smooth meadow grass seeds early, beginning in early May. Its seed is very fine. This species is very hardy and can tolerate drought well. Smooth meadow grass forms a dense sward but germinates very slowly, and spring growth is also limited. Its taste is less acceptable.

Tall fescue

Tall fescue does well in both dry and wet conditions. Acceptability is not a good as perennial ryegrass, so cows consume less.

Moderate grasses

Meadow foxtail

The flower of meadow foxtail is an ear that resembles timothy grass. The plant flowers as early as April. It is not an attractive option for grazing, as the grass becomes less acceptable with age.

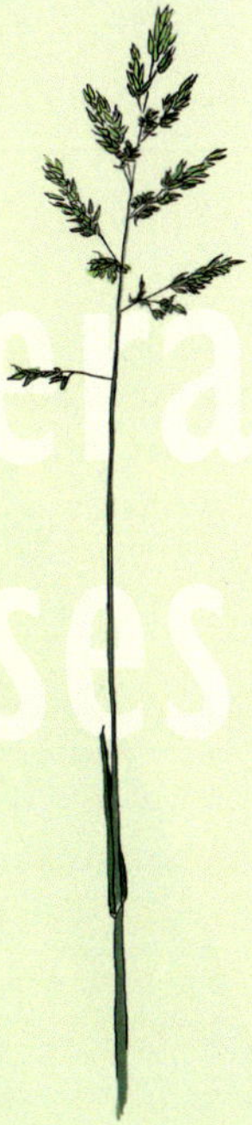

Yorkshire fog

Yorkshire fog is very easily recognised by its violet stripes ('pyjama bottoms') on the sheath. This species prefers damp, acid sandy and peaty soils. Yorkshire fog is not acceptable to livestock, amongst other things due to its hairy leaves. Grasslands with a lot of yorkshire fog can have problems with clumping.

Creeping bent grass

Creeping bent grass flowers as a wide triangular plume. It sometimes forms very long stolons above the ground. This is one of the less desirable grasses as its nutritional value is not high. Creeping bent grass spreads out across grassland, especially in autumn, making it highly competitive.

Rough stalked meadow grass

Rough stalked meadow grass in bloom looks much like smooth meadow grass. However, rough stalked meadow grass has many more limp tillers, and likes moist ground more than smooth meadow grass. The dry-matter yield of rough stalked meadow grass is considerably lower than that of perennial ryegrass, mainly due to limited growth in June-July.

Couch grass

The ears of couch grass are perpendicular to the axis, making it look different from perennial and Italian ryegrass. Couch grass reproduces mostly by very fast-growing stolons, and can easily crowd out other grasses on fertile soils. Couch grass does not tolerate very frequent short cuts and shade.

Bad grasses

Annual meadow grass

As its name implies, annual meadow grass is an annual species that flowers throughout the year. Since soil often contains so many of its seeds, it quickly fills up bald patches. It does not withstand heat or drought well.

Marsh or bent foxtail

Marsh or bent foxtail is a fairly small perennial species with a noticeable curve in the stalk. This species flowers from May to August. It has flowering plumes with ears 2.5 to 3.5 mm long. Marsh foxtail grows mainly on moist meadows, principally alongside ditches where the pH is low.

Grass-clover

Clover in the grassland fits in well with most farms. The advantages of clover are its ability to fix nitrogen and its good palatability when a reasonable amount of white clover is mixed with grass. In association with bacteria, clover fixes nitrogen from the air; this can be as much as 200 kg N per ha. The less nitrogen from manure or artificial fertilisers you use, the more actively clover will fix nitrogen. This makes clover mostly useful when the N-fertilisation level is no higher than 250 kg per hectare and the soil's ability to provide nitrogen is low. The term 'grass-clover' usually means grass with white clover. Red clover is often overlooked – unjustly so, because unlike white clover, red clover grows throughout the year. There are two ways of introducing grass-clover mixes onto a dairy farm:

1. Grass with red and white clover, purely for cutting.
2. Grass with white clover for a mixed operation.

Estimating clover content

Estimating the clover content in the dry matter takes some practice. What looks like a lot of clover usually works out at a limited amount in the dry matter. Generally, the dry-matter amount is half of the visually estimated amount.

1. 5 to 10 percent
2. 10 to 20 percent
3. 20 to 35 percent
4. 35 to 55 percent

White clover

White clover is the most commonly used clover, since it withstands grazing as well as mowing and does not disappear too easily from the sward. White clover has a high nutritional value and when combined with grass, provides a readily eaten forage for cows. A clover content of 30 to 50 percent of the total dry matter in summer is optimum for silage/grazing fields. Swards managed purely for silage can contain up to 60 percent. At a higher concentration of clover, the protein and total dry-matter yields of a field will decline. If there is too little clover, too little nitrogen will be fixed and yield and nutritional value will be less.

The proportion of clover in a sward varies greatly over the year. It is low in spring, rises gradually to a peak in summer, and then declines again.

Managing the proportion of white clover

There are a number of measures you can take to influence the amount of white clover in the sward. The quantity of the first cut has a great deal of influence. A heavy cut will inhibit the amount of white clover. Grazing the first cut, or light mowing, will stimulate it. Since clover needs higher temperatures to start growing than grass, its development in spring starts later than grass. A light cut means a decrease in competition from grass, stimulating clover growth. The cutting height also has an influence. Mowing short with a stubble length of 4 to 5 cm instead of 7 to 8 cm promotes a higher proportion of clover in the grass. This is due to grass's reduced ability to re-grow from a short stubble, and the low cut also means the clover plants get more sunlight.

▲ *Clover is more vulnerable on compacted soil than grass. This is because clover has thicker roots than grass, and has more difficulty penetrating the soil.*

◄ *White clover can spread across a field using stolons (horizontal stems). A petiole (leaf stalk) with a leaf will grow up from each bud. Each bud also has two root growth points, one on each side. A single clover plant can extend 20 to 40 cm 'around itself' in a year.*

Alternating mowing and grazing contributes to an even distribution of clover across the field. Frequent consecutive grazing can result in tufting, in which clover does less well. ▼

Prussic Acid (hydrogen cyanide)

White clover contains the substance cyanoglucoside, which can form highly toxic cyanide. In theory, cyanide from clover plants can result in delayed growth and alter the taste and smell of the milk. However, in practice these signals have not been observed under Dutch livestock practices.

Horses can develop health problems from cyanide formed in white clover, however, such as a slower gait in the hindquarters, bladder paralysis and reduced fertility.

White clover suffers less from drought than grass does at first. This is because clover is less dependent on water for nitrogen transport in the soil. If drought persists, however, clover will die sooner than grass because of its less extensive root system. ▶

Carrying capacity

With a high proportion of clover the soil's ability to take weight decreases due to less soil organic matter being produced. This is mainly a problem during wet periods. For this reason, aim for a balanced proportion of clover development throughout the year.

▲ *If white clover is in full bloom, the chance of bloat in ruminants is reduced.*

Bloat

Eating large quantities of clover can cause health problems in cattle. The easily soluble proteins in clover can cause froth to form in the rumen. The gas bubbles are trapped, causing bloat. This risk is greatest with a high percentage of clover on a field with young grass that still has little effective fibre, and if the cows are turned out onto the field hungry. Cows that eat other fodder besides grass-clover, preferably with a high energy and fibre content, are seldom troubled by bloat.

Herbs with a high tannin content, such as birdsfoot trefoil (*Lotus corniculatus*), can help reduce the possibility of bloat in grass-clover swards.

Which clover do you choose?

Characteristic	*Red clover*	*White clover*
Life span	2-4 years	perennial
Reproduction	seed	seed and stolons
Management	cutting only	cutting and grazing
Digestion coefficient	lower efficiency	higher efficiency
Spring growth	good	moderate
Productivity	11–15 tonnes dry matter	10–12 tonnes dry matter
Drought resistance	good	reasonable

Red clover

One essential difference between red and white clover is that red clover does not form stolons, and therefore cannot spread vegetatively through the sward. Thus, the areas in the fields containing no red clover will never be colonised by it. Since clover plants only live for a few years at most, red clover will slowly disappear from a field, unless it's given the chance to flower and set seed.

Frequent grazing of grassland will cause red clover to disappear quickly, since its root reserves are soon exhausted.

This makes red clover more suitable for silage cuts for one to four years. You can of course seed a combined-use field with red clover for a higher dry-matter yield in the first year.

▲ *It is difficult to manage the proportion of red clover. Unlike white clover, red clover benefits from heavy cuts. After several heavy cuts, red clover can get the upper hand in the sward.*

◄ *Red clover has a substantial taproot which can go very deep into the ground. This makes the species a real pioneer plant that grows even in less fertile soil. Red clover does well on drought-prone ground, but can also maintain itself during wet periods.*

Allowing red clover to flower and form seeds is one way to ensure the plant does not disappear from a sward after a few years. During the second and third cuts in a good summer, red clover can form a lot of seed which remains viable for years. ▼

Clearing tufts

Cows do not graze evenly over a paddock. They leave more grass in some places than in others, which creates 'tussocks'. Mow the tussocks with a mower or topper after two grazings. If you intend to make the next cut a silage cut, you can leave the tussocks if you like. Remember that this will affect the quality and taste of the ensiled grass, however. If the silage cut is meant for winter fodder for milking cows, it's better to clear the tussocks first.

Grazing after the first cut

With grazing in mind, it's better not to take all the first cut in one go. Differences in re-growth rates often help you to have paddocks with different amounts of grass. Turn the cows into the tallest paddocks for a shorter period than you would otherwise, and feed them more at the trough.

If you do take the first cut in one go, in the following weeks there won't be any paddocks available in differing stages of growth. In this situation as well, modern set stocking may be a solution.

After a week, add enough paddocks together to allow the cows to graze for about two weeks, combined with supplementary forage at the trough. After two weeks, the cows can be turned into a paddock with more grass, where they can graze for three to four weeks.

It's possible to keep working in this way for the rest of the grazing season. An alternative is to switch to a form of rotational grazing, such as restricted grazing. This can yield 15 percent more grass per year than with set stocking.

◂ *Different grazing systems can be applied in sequence during the grazing season. The months following the first cut are very suitable for day-and-night grazing, since you're letting the cows eat as much as possible of cheaper meadow grass.*

The best point at which to turn cows out onto rotationally-stocked pasture is when you have herbage mass of 3000 to 3200 kg dm/ha. The grass will be about 15 cm long, about as long as a fist with the thumb raised. ▸

Planning and record-keeping

Optimising grassland utilisation means keeping a lot of information. This can be done on paper as well as with convenient computer software. You'll need to record data about use, yield and fertiliser to adjust and improve grassland use in the future. For example, if you'd like to be able to apply fertiliser as needed, you have to have some idea about what you intend to do with a paddock. Combining an appropriate grazing schedule with alternating cutting and grazing requires good planning.

Practice shows that it's difficult to adhere to a plan. Weather conditions in particular can seriously disrupt your grassland management plans. However, that is not a reason not to make a plan. Your plan can even stay in your head – it's easier to make a change there than sitting at the computer.

Siesta grazing

With siesta grazing, the cows go back to the yard at midday, where they eat low-protein fodder such as maize silage. In the evening after milking, the cows go back outside for a few hours. An advantage of siesta grazing is better nutrient utilisation, which increases milk production. Limiting the hours of pasture access may also improve nitrogen utilisation. A major disadvantage of siesta grazing is the extra work bringing the animals back in. This is the reason why so few dairy farmers use siesta grazing.

A robotic fencer can automate the moving of movable fence-lines. These little carts ensure a strip of fresh forage for the cows each day. The advantage is a constant quality of grass and a calmer herd. ▼

A grassland use calendar is a helpful tool in managing grassland. The calendar makes it simple to track the use and fertiliser applied to the various paddocks. It is also handy for planning grassland use and for modifying the plans. ▼

If possible, graze calves on aftermath, giving them a fresh paddock every two weeks. This helps keep the infectious challenge from lungworm and intestinal worms low. This allows the animals to build up their resistance without becoming ill themselves. ▶

▲ *After the second cut, growth slows down and more paddocks will be needed for grazing. The area to be mown decreases after each consecutive cut and the cuts become somewhat lighter, at 2.5 to 3 tonnes per hectare.*

Cutting in summer

The growth rate of grass declines after mid-June, hence it takes longer to make an adequate silage cut in the field. At the same time there is a very good chance the grass may flower. Once this happens, the total nutritional value yield of a paddock will not increase and the nutritional value per kg dry matter will decrease. Cut a paddock as soon as you see the first signs of flowering.

Another problem common in June and July – and in the months thereafter – is dry weather. Dry weather can really make your grassland planning go awry. It will halt growth, and even a small harvest will be full of stems. It may then be necessary to cut before an adequate silage cut is presented in the field.

More protein after dry weather

After a dry spell in summer, grass usually does not need extra nitrogen to stimulate growth. Nitrogen from earlier applications of fertiliser has not been used, and as a result of mineralization from organic matter, additional nitrogen is released when it rains. The grass that grows after this rain is usually high in protein. Bear this in mind when giving supplementary feed to the cows.

High-fibre grass

Cows need enough fibre, or roughage, in their diet for the rumen to function properly. Grass with lots of stems (high in crude cellulose) has a high fibre value. An increase in fibre value of grass correlates with a decrease in digestion coefficient and thereby in nutritional value. But if your milking cows are given forage with plenty of low-structure products whilst housed, it may be desirable to add grass silage with a high structural value. Grass with a high fibre content can be harvested in the early summer, as long as the grass has still to flower. The timing of the silage cut is decisive: you want to balance the increase in stemminess when grasses begin to flower with the decrease in nutritional value from going to seed. When the weather is good and there's enough of a crop, but you still have your doubts, opt for higher nutritional value and not for stems. Re-growth after taking a silage cut that has headed can be rather disappointing, with an open sward as a result.

Puzzler

The choice: do you graze or cut?

This field had two consecutive grazings. This has caused a lot of tufting. The paddock should now be mown. Mowing as service to grazing is the best method to retain 'clean' grazing paddocks. Remember: 2 x grazed = silage cut.

◄ *Festulolium is a cross between ryegrasses and meadow or tall fescue. This grass lends itself well to making silage with extra structure. There are also dairy farmers who simply use tall fescue for producing high-structure grass. An advantage of tall fescue over ryegrass is that its nutritional value decreases less as stem content increases.*

Puzzler

What happened here?

There are visible bands of shorter grass in this pasture. What could have caused this? The bands indicate that the soil was compacted by heavy machinery. This field was probably driven on at a bad time, and with too high a tyre pressure.

Manuring during summer

Manuring during the summer months is mainly a matter of adjusting fertiliser to maintain grass yield and quality (crude protein). When applying manure, especially the weather and grazing play an important part. Do not apply slurry on a field which will be grazed directly afterwards. In the short term, manure/slurry will adversely affect the acceptability of grass. Try to apply slurry on a field when the next cut is going to be a silage cut.

Always take a critical look at grassland use when applying slurry. The loss of potash and phosphate is greater on fields that are cut more often for silage than being grazed. Bear this in mind when applying slurry. Give fields that you're intending to cut for silage extra slurry to provide the potash and phosphate they need.

If you have enough storage capacity, early August is a good period to stop applying slurry. You can make the best use of the organically bound nitrogen in slurry, which will remain active for months, and avoid losses.

Apply the last slurry on the land before mid-August if your storage capacity allows. ▼

Cow pats and urine spots

When cattle graze, their waste does not end up in the manure pile, but goes directly onto the pasture. This saves you the expense of spreading it, but you will have an uneven distribution of manure and urine across the paddock, and thus less than ideal manure utilisation. To prevent nitrate leaching on sandy soils, it's important to get as much manure and urine in the slurry store.

▲ *When combined with sunny weather, urine with a high nitrogen content can seriously burn the grass, as in this photo. In recent years, urine spots have become less common. This is due to the lower protein content in the ration because of, amongst other things, reduced nitrogen application.*

▲ *Cutting paddocks that have been grazed can result in the grass being contaminated by cowpats, especially when you mow after a wet spell. This will adversely affect the taste and finish up in the machinery. If you can, ensile grass from grazed paddocks separately, so that you can use it for the younger animals and dry cows.*

▲ *In general, a diet low in protein and high in fibre will result in more 'soil-friendly' manure. This means manure in which the soil organisms can become established and break it down quickly. A soil-friendly cowpat will quickly disappear.*

Since this cowpat is full of holes, you may think that the soil organisms are busy breaking down the manure. But looks can be deceiving. It's been attacked by crows looking for maize grains in it. ▶

▲ *Many factors will determine whether irrigation is economically feasible.*

Irrigate or not?

A lack of moisture is the main reason for reduced grass growth during the summer. Dry weather not only reduces the yield but contributes to a decrease in pasture composition and quality, especially on sandy soils vulnerable to drought. Poor growth gives weeds an opportunity and the drought-resistant poor grasses, principally couch grass, can thrive. You can prevent or limit damage from dry weather by irrigation. You can also irrigate to ensure you don't run out of grass for grazing. However, irrigation is expensive. It involves investing in the installation, possibly drilling for a source of groundwater, laying underground pipes and consuming energy.

Irrigation is also labour intensive, and you exert pressure on the supply of ground- or surface water. All these are good reasons to really weigh up each alternative to see whether it's a sensible solution. The government and local water authorities may also limit how much water you can abstract, which may also affect your decision. Compare the costs of irrigation and the costs of forage from the economic perspective. In an extremely dry year, feed prices rise sharply. Having an irrigation system can be seen as an insurance policy for dry years. Whether you need one will depend on how vulnerable your pastures are to drought.

If there is a lot of forage available, irrigation is less attractive. ▼

Meadow grass is a signal that you've over-watered. It will disappear in dry periods from areas that are not irrigated. ▶

Correct time?

Look at the ground, not the sky

Start irrigating at the right time: the point when there's a lack of moisture in the part of the soil where the grass roots are. There are various methods for determining when to turn on the irrigators. There are computer programmes and 'irrigation guides' with which you can calculate when it's necessary to commence irrigation on the basis of soil type, precipitation and evaporation. In several Dutch counties, the use of the irrigation guide is compulsory to avoid wasting water.

Assessing the need of each area is a more practical method than using a computer programme. You can take soil samples from the root zone at various areas in a field using a soil sampler. Kneading the ground in your hand will give you an idea of its moisture content. If you want an even better estimate, assess the groundwater level and root depth as well. The groundwater level can be determined by drilling a hole with a soil drill and measuring how deep the water is. The root zone is measured by digging holes.

Depending on the subsoil, the capillary rise of the groundwater will be between 30 and 100 cm.

▲ *How much water to apply each time depends on the root depth. The greater the depth, the more losses are reduced following a heavy dose of water. It's usually wise not to irrigate more than 25 mm at a time.*

The soil is easy to knead, and holds its ball shape. It doesn't need irrigation yet. ▼

◀ *Most farms have areas where the soil dries faster, such as higher-lying areas in the fields. Keep an eye on these signal areas as soon as a moisture shortage threatens. If the grass begins to wilt over the course of the day, it's a sign you'll need to irrigate.*

▲ *Grass from environmental areas is often baled. This is a flexible way to conserve ensiled grass. An advantage of baling this type of grass is that it's easily combined with grass from conventional silage clamps. Mixed ensiling is also a possibility.*

Environmental grassland

Sometimes grassland has a wider function than purely providing feed for animals. When natural values and protection of meadow birds play a part, or even as the main objective of management, the task of management will look very different. Grassland with this purpose has greater biodiversity. Generally, roughage from grassland in environmental areas can be combined well with roughage from conventional meadows for livestock purposes. The lower nutritional value is offset by a higher structural value, which stimulates the action of the rumen (scratch factor in the rumen).

Postpone cutting date

Taking cuts from environmental reserves is often not allowed before a specified date. Cutting late in the season usually means the first cut will be sizeable. Moist environmental grass is difficult to conserve and yields silage that cows dislike. It's preferable to mow following a week of drying weather, when the subsoil will be dry and the grass will remain smelling fresh.

◄ *Grass from environmental areas has the highest nutritional value if you cut before it goes to seed. Has it already headed? Wait for good weather before cutting. At this point it doesn't make a lot of difference whether you wait a few days or a week.*

It's all well and good to provide high roughage forage, but it has to get eaten too. Cows can't be choosy when you're using a feeder mixer wagon, but if you offer high-roughage forage separately, it will have to appeal to them. ►

Grassland composition

Management decisions influence the composition of the plant species in an environmental area. If the land was previously used for agriculture, the primary objective will often be to improve poor soil quality. Methods available are fertilising and mowing or grazing. If the objectives include protection of meadow birds, using fresh farm yard manure will be desirable as the birds will use the straw in it for nesting material. However, it will not automatically result in more worms in the soil. Cutting the first cut as soon as possible after the earliest date permitted has a major influence on whether or not there will be loss of biodiversity and a takeover by weeds. Postponing the cutting date often quickly results in dominance of yorkshire fog, since this species rapidly goes to seed. By alternating cutting and grazing, you can guarantee a relatively good and long-lasting sward.

Tansy (top) and St John's wort (bottom) are often mistaken for ragwort. ▼

In order to reduce high phosphate levels in the soil in environmental areas, it is useful to sow a mix of grass and white clover. If you also use potash fertiliser, you can achieve a good dry-matter production whilst getting rid of excess phosphate. ▼

Toxic ragwort

Ragwort is a highly toxic plant which is sometimes abundant in environmental areas. The alkaloids it contains can cause liver damage in mammals, which can be fatal. Horses and cattle are especially at risk from ragwort. Sheep have enzymes in their digestive tracts which render the alkaloids less harmful. Grazing animals stay away from this unappetising plant in the pasture, but when it's dried, they cannot distinguish it from the rest of the silage or hay, and eat it unawares.

If there is only a bit of ragwort, it can be removed manually. Do wear gloves, because the toxins can get on your hands and cause a rash. A silage cut from fields with 5 percent or more ragwort is not suitable for feeding to livestock.

After a growth rate reduction in July, grass often has another growth spurt in August-September, once the soil moisture content improves. Late summer is also a good time to sow grass and clover. However, with the onset of autumn, nuisances such as rust and insect damage also present themselves.

▲ *When choosing grasses for re-seeding, bear in mind their vulnerability to crown rust. The variety with the lowest resistance determines the vulnerability of the total mix.*

Preventing crown rust

Crown rust is the only fungal disease that causes large-scale problems on grassland. The fungus, *Puccinia coronata*, often strikes in August-September. Crown rust is easily recognised as small orange-yellow pustules on the grass leaves. If you see rust on grass stems, it's probably black stem rust. A crown rust infestation gives a field of grass an orange-yellow glow. At the point when this is visible, the plants have already had the disease for about ten days.

Crown rust takes its opportunity in the autumn, when the grass is stressed from such factors as drought or a lack of nitrogen or potash. It gives grass an unpleasant taste and cows will avoid it. Using a fungicide is too expensive, however; the best measure to prevent further spread and damage is to cut and ensile as soon as possible.

Crown rust is less quick to rear its head on fields with clover, since in late summer the clover will be providing a steady supply of nitrogen. On fields without clover, it's important to choose varieties of perennial ryegrass with high crown rust resistance.

Other fungi

Besides crown rust, there are other fungi which can cause incidental damage to grass. Leaf spot, recognisable by small watery brown spots on the blades, is the result of fungi of the genus *Drechslera*. Once the grass has been cut these fungi usually do not grow on the new grass that grows. Footrot, caused by the *Fusarium nivalis* fungus, occurs mainly from autumn to spring. A serious infestation can cause major damage to the sward as plants will die. Footrot is infamous for spreading under a layer of snow.

Silage affected by fungal infections can also release toxins, the so-called mycotoxins. *Mycotoxin zearalenone* (ZEA), in particular, is often found in silage. ZEA is the product of a fusarium fungus. Cattle seem to be less sensitive to a small amount of mycotoxins in their diet, due to among other things the buffering action within their rumen. Other animals such as goats are more sensitive, so do not feed them grass or silage with any fungal contamination.

▲ *Small slugs can cause a lot of damage, especially to clover. They leave behind a recognisable pattern (photo). The pea leaf weevil (*Sitona lineatus*) is also a fan of clover leaves. You can minimise the risk of damage from snails, slugs and beetles by choosing an appropriate clover variety.*

◀ *Snails can destroy many germinating clover seedlings when you re-seed, even if you've chosen a variety less susceptible to snail damage. Rolling the seedbed firmly can help prevent damage, as the snails can less easily find shelter from sunlight.*

Crane fly

Crane flies lay their eggs in the grass from August until early October. The larvae that emerge are called leatherjackets. They mostly live underground, but feed on the parts of plants above ground, causing a lot of damage. Leatherjackets seem to prefer dicotyledonous plants to grasses, which may even mean they can help keep weeds out of the pasture.

Mice

In some years, mice occur in very high numbers. In these 'mouse years' the rodents can cause a lot of damage on pastureland. The culprits are mainly field mice and bank voles. Both species eat a variety of foods. They can gnaw at roots, and they dig a great deal. If you have them, you won't get normal grass growth. Mice can be an enormous nuisance, especially on sandy and peaty soils. There is virtually no way to get rid of a mouse infestation. If you do have damage, rolling the sward is useful to repair the connection to the subsoil.

Pasture can be destroyed by mice during 'mouse years'. There is actually no effective remedy against mice. ▼

Hares love to spend their time on pastures, and they love to eat grass. However, their numbers are so low that they cause negligible damage. ▼

Swans prefer newly seeded grassland and can cause a lot of damage by devouring, trampling and smearing grass with excrement. ▼

Grubs

Grubs are the larvae of a group of beetles. Grubs living under grassland are mainly those of the May bug (or cockchafer) and garden chafer or rose beetle. They can cause a lot of local damage, particularly on high, dry sandy soil. In the summer months they devour grass roots. This loosens the sward and the grass dries out. The loose and rolled-up sward makes injecting slurry difficult. Additional damage can be caused by crows, hedgehogs, foxes, and badgers which tear open the sward to get at the grubs. Intensive grazing is beneficial as it consolidates the sward.

If damage is minor, you can let the sward root again by rolling and watering it. More serious cases will need re-seeding. Grubs can be biologically destroyed by spreading parasitic worms on affected areas. This is expensive, however, and only works if you know precisely the type of grubs you're dealing with and the appropriate species that attacks them.

No chemicals

Using chemicals to kill grubs is not legally permitted. Grubs are often confused with leatherjackets. One major external difference is that grubs have legs and leatherjackets do not. Grubs are also much thinner.

Wild carrot or Queen Anne's lace is food for the garden chafer dagger wasp, which preys on grubs of the rose beetle. Planting it, for example along field margins, can help prevent damage from grubs. ▼

Grub damage is mostly seen on poor sandy soils near woods. If damage is minor the sward can form new roots. Use a roller if needed. ▶

The sod, in seven steps

1. Dig out a 20 x 20 cm clod with a shovel, down to 30 cm depth in one piece.

2. Supporting the sod, bring it up and lay it on the grass.

3. See if you can distinguish different layers, and measure the thickness of each.

4. Assess if there is root growth in each layer. This way you can localise problem layers.

5. If you've dug up the sod from moist soil after a dry period, then you'll see white root tips. They make it clear where the root activity is. The shallower the root depth, the sooner you should think of re-seeding.

6. Assess how crumbly the soil is in each layer, and estimate the percentage of soil crumbs in relation to the total soil volume. If the soil has good structure, you'll see:

   - 100 percent crumbs at 0 to 5 cm deep;
   - 50 percent crumbs at 5 to 10 cm deep;
   - 10 percent crumbs at 10 to 30 cm deep.

7. Assess the activity of soil organisms in each layer. Count the worms you can see. Four or more in a 20 x 20 cm clod indicates healthy soil activity.

Examining a sod of soil

Take a look below the surface in summer too. Especially if you're considering re-seeding or rotating crops, it's important to look beyond the botanical composition of the pasture. It's not always necessary to dig a pit.

Evaluating a sod of soil you've dug up with a shovel can provide a wealth of information. It is quick too, so you can look at several areas in a field.

Remove a sod by loosening the soil with the shovel from side to side.

Carefully remove the sod, keeping it as much as possible in one piece.

Measure the root depth.

Evaluate how crumbly each layer is.

Improving the soil

August is often the right month to make improvements in the soil, such as levelling and loosening problem layers before sowing new grass.

You can delay the sowing of new pasture on sandy and chalky soils to the early autumn by sowing an intercrop. Cereal crops lend themselves well to this, as they root deep and improve soil structure. Since cereals can be harvested in July-August you'll have enough time for the soil improvement activities. Cereals deplete the soil of nitrogen, which is good for establishing the clover in a grass-clover mix.

◄ *Using a subsoiler to lift problem layers up from below often only has a temporary effect on compacted layers. Loosening up the soil, however, can achieve a long-term improved soil organism activity and better root growth. Common earthworms, for example, can bore through plough pans.*

In August the weather is often dry, so it's a good time to work on the land. Improvements such as cultivation and levelling are often best done in August. ▼

Subsoiler attachments on the plough will help to prevent plough pans. ▼

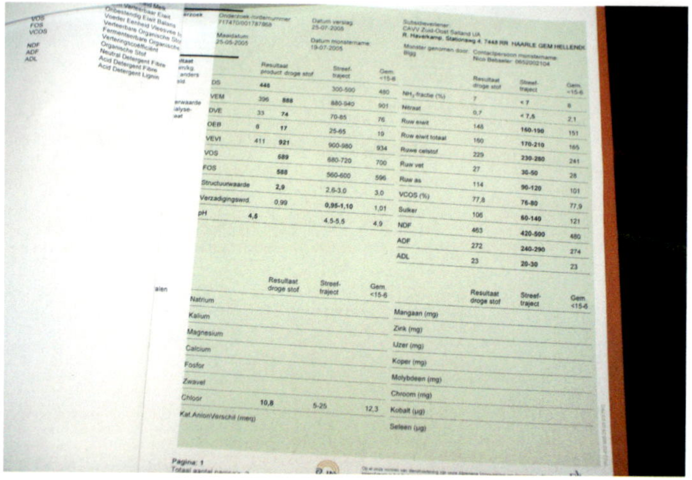

▲ *Silage analysis gives information about decreasing grass quality. The most important thing to watch out for is the digestibility of the organic matter of the dry matter (d-value); compare this with the reference values of silage in the same season.*

To sow or not to sow?

The decision whether or not to re-seed grassland will depend on various factors. Besides the state of the soil and desirability of taking measures to level out and/or drain it, the primary role is the composition of the grassland.

Wageningen university has developed a programme called the 'Herinzaaiwijzer' (in Dutch, www.asg.wur.nl under 'Producten') for calculating whether re-seeding is financially sound or not. It calculates the difference between the cost of making improvements and the expected extra yield of a pasture in the first five years after re-seeding.

Use as many of the signals you've received during the year from the grassland as possible when deciding. For example, if the proportion of couch grass or meadow grass has increased sharply, when and where puddles appear on the land after rain, or if you now have more moisture-loving grasses such as meadow foxtail.

Clover is more likely to establish well when sown after an intercrop has been grown on ploughed pasture. An intercrop not only provides a soil low in nitrogen, but fewer clover cyst nematodes as well. These pests can stunt the growth of or even kill clover seedlings. ▼

Late summer is a good time for mechanical weeding. By creating a false seed-bed you can eliminate many weeds before you sow grass-clover. ▼

Use decides the blend

When sowing grass, you must have a good idea of how you'll be using the fields concerned. This will largely determine the best grass and clover types to sow. Broadly speaking, there are two types of use:

1. Alternating grazing and cutting.
2. Mainly cutting.

Most grass seed is sold in blends containing several grass species and/or varieties, and sometimes clover as well. In choosing a blend, besides the use planned (mainly grazing or cutting), the intensity of use and the nitrogen fertilisation level are also important to consider. Most blends are based on perennial ryegrass. A blend can contain both diploid and tetraploid variants of perennial ryegrass.

▲ *Tetraploid ryegrasses (right) have twice as many chromosomes as diploid varieties and usually somewhat wider blades. They are also slightly darker in colour. The sward is more open with tetraploid grasses than with diploid grasses. Grass seed mixes contain diploid or tetraploid grasses or a mixture of both. Since tetraploid seeds are larger than diploid seeds, more of them need to be sown to ensure the same sowing density.*

Choosing a blend

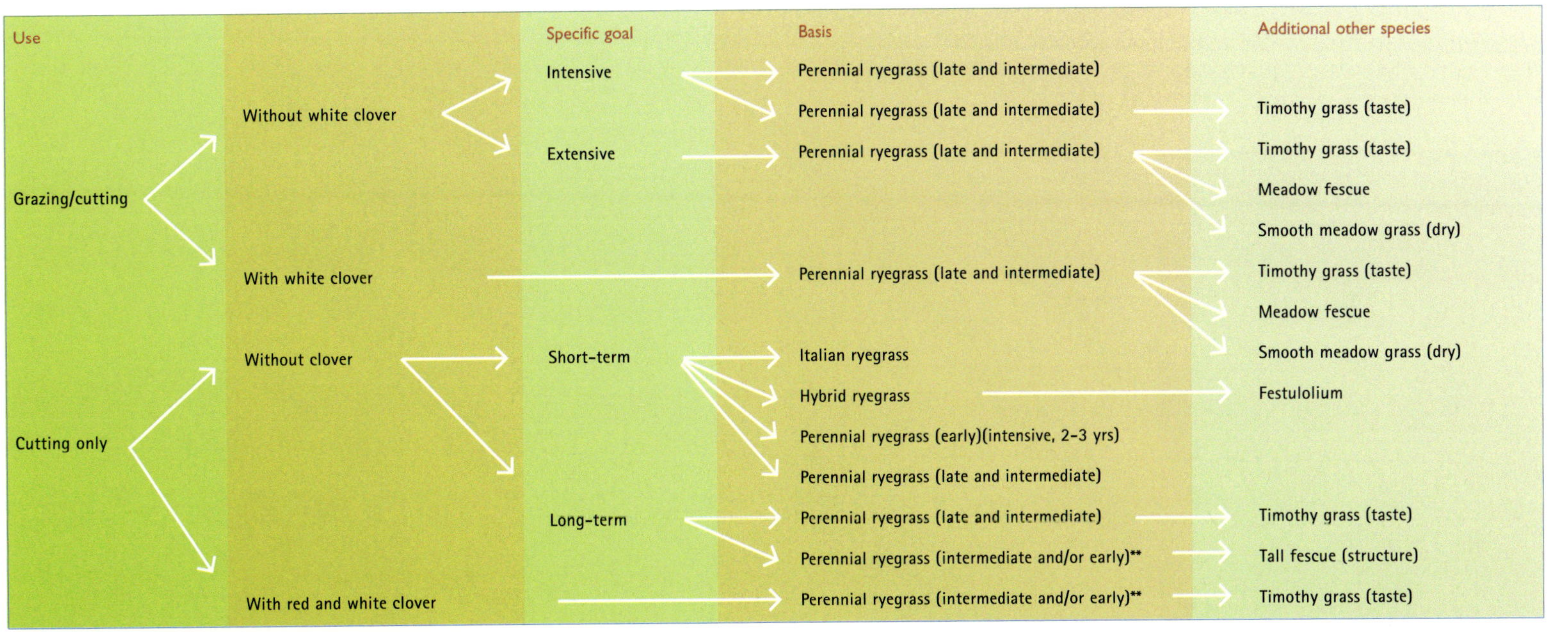

* *with tetraploid perennial ryegrass

** Intermediate and early perennial ryegrass has more structure than late

▲ *The nutritional value of red clover is not much less than that of white. However nutritional value assessments usually show lower energy and digestibility values. This is because the current analysis and calculation methods underestimate the nutritional value of red clover.*

Mowing? Red clover!

Red clover is highly suited to use in fields that will be cut for silage over several years. A grass-clover field using red clover will produce several more tonnes of dry matter than with white clover per year. When sowing a field for mowing, it's better to sow both red and white clover in addition to the grass. The red clover will last from two to four years, depending on the variety, and afterwards the white clover will invade the areas left open by the red clover.

Red clover is highly productive, but encounters some resistance from livestock farmers since it can be tough, and cows may avoid it. If you chop it, or ensile it after chopping it up, you won't have a problem.

Grazing? White clover!

If pasture is intended for mixed use, with mainly grazing, white clover is recommended, since red clover won't last for long on fields that are also regularly grazed. For a quick start to production, it can also be a good idea to add two kilos of red clover per hectare on fields for mixed use.

Just plough, do it now!

Seeding clover on existing grass often gives disappointing results. If you want clover in your sward, it's really best to get on with it and plough. Only farms converting to organic production methods should over-seed, since it's not feasible to improve all the pastures at once.

A light slurry application of 20 m³ per hectare before seeding will help ensure a new grass-clover crop receives a good start. ▼

Grass-clover recipe

Sow grass-clover before 1 September. Sowing later reduces the chances that the clover will 'take'. If you have to sow on or after 1 September, use one or two additional kilograms of clover seed per hectare.

For silage

For a silage field use:

- 25 to 35 kg grass seed;
- 5 to 6 kg red clover seed;
- 3 kg white clover seed.

In the UK, consult the NIAB list (membership required) to choose the appropriate varieties of grass and clover for you (see www.niab.com). In the Netherlands, use the Ministry of Agriculture's list (see www.rassenlijstlandbouw.nl).

Italian ryegrass should not be combined with clover. It often competes too strongly with clover, which never has the opportunity to develop. Mixtures with hybrid ryegrass are useful.

For mowing and grazing

Use in mixed-use fields:

- 25 to 35 kg grass seed;
- 3 to 5 kg white clover seed;
- Optionally 2 kg red clover seed.

▲ *Get rid of couch grass and docks thoroughly before re-seeding. This is best done using glyphosate for couch grass and MCPP for docks. A mechanical approach is also an option, with a Kvik-up machine (see photo), which removes weeds and lays them on the surface. The roots will dry out on the surface, and later the dead material will be ploughed back in.*

Adding herbs

Other herbs besides clover have useful qualities and can be combined with grass. The most suitable are chicory (cv. Puna or Commander) and plantain. Use 2 kg seed of either type. Chicory helps by adding minerals to the animals' diet, including calcium, sodium, phosphate, copper, cobalt, selenium and sulphur. Plantain is good for the animals' health due to the antibacterial substances it contains. Chicory and Plantain do not lower the productivity of the field, and even increase production during dry weather.

◄ *If your crop - rooting – soil organism – soil cycle isn't functioning, you may have no other choice than to plough and re-seed the field. Cultivating the soil means the grass can root deeper (see photo) which attracts more soil organisms and improves the soil structure.*

Fresh white roots

Young plant roots are best able to take up moisture and nutrients from the soil. A continuous new growth of young, active roots (white in colour) is necessary for good production. Moreover, grass roots exert a beneficial influence on the soil structure. A healthy sward is therefore your best instrument for improving the soil.

When re-seeding, what's important is to get the crop - rooting - soil organism – soil cycle established as quickly as possible. It can help to add fresh organic matter through a green manure crop or in the form of fresh farm yard manure. Growing a crop of cereals before sowing grass (or grass-clover) is beneficial. The roots of cereal plants introduce organic matter lower down into the soil structure, and literally and figuratively open the route for good root growth after the grass seed has been sown.

Puzzler

Clover or chickweed?

Seedlings of clover and chickweed are often confused. Clover is in the grass in the photo on the left and chickweed on the right.

Clover works with rhizobium bacteria to fix nitrogen from the air. These bacteria are almost always present in the soil. On contact with clover roots the bacteria are enclosed, forming nodules. The pink colour indicates healthy activity. ▼

Signal plants

A closer look at the other plants in the sward can provide a lot of information. Yarrow and sheep sorrel are for example indicators of dry grassland. If you keep your eyes open, these grassland indicators can act as business advisers. Immediately after sowing, the composition will generally be a reflection of the seed blend. It will take several years before you get usable signals from the botanical composition. When plants occur in patches in a sward, it points to an unbalanced or not yet stabilised situation.

Plants that indicate wet conditions can also indicate soil compaction. Examples are marsh or bent foxtail, water manna grass, reed manna grass, reed canary grass and water-pepper.

▲ *Yarrow, an indicator of dry environments.*

▲ *Reed manna grass is a coarse grass. Its stalks can reach a centimetre in thickness. Since it prefers marshes or marshy ground, it is an indicator of wet conditions.*

▲ *Marsh or bent foxtail is an indicator of wet grassland. This grass species grows in wet meadows. It is mostly seen along channels and ditches where the lime content is low.*

Knotweed often grows in spots with a lot of cow traffic, such as the gateways into fields and on the edges of the cow track. ▼

Plantain indicates a good supply of phosphate: this plant reacts more quickly to phosphate than perennial ryegrass. ▼

Reed canary grass loves nitrogen and water. This grass is often seen on marshy ground in wet meadows and on the banks of ditches, rivers and ponds. ▼

Grazing during late summer

When grass growth slows in August, it's better to turn the cows into shorter swards, so they aren't forced to eat older grass. Assume a herbage mass of 2800 to 3000 kg dm/ha instead of 3200 kg dm/ha. The maximum age desired for grazing grass is four weeks, and for silage cut six weeks.

Clover

In late summer, the clover content can increase substantially. The result is that grazing cows take in high-protein forage. Bear this in mind when feeding the animals at the trough. The urea content of the milk is a helpful indicator for directing the protein-energy ratio in the total food supply.

Nutritional value

Starting in August the energy value of fresh grass gradually declines. This is one reason to confine calves indoors in early September, partly because grass becomes less acceptable as autumn approaches, and intake decreases.

Topping or clearing?

As the summer progresses, maintaining the acceptability of grass will demand more of your attention. It can help to remove tufts and remaining plants with a topper, disc or drum mower or a flail topper. If you use a conventional topper, it leaves a lot of leftover grass on top of the sward, which has an unpleasant taste, especially in wet weather. This is less the case when you use a flail topper. This is why cows prefer to graze on a flailed sward rather than a topped one.

Dry weather is detrimental to grass growth. An advantage of dry weather on clay soil is that it improves the soil structure. It has a greater structure-improving effect than frost. ▼

The crude protein content of a grass-clover crop can rise sharply in the summer and autumn. ▼

Mowing during late summer

If possible, late summer and autumn cutting should be even more of a service to grazing than in previous months. A paddock needs to be cut regularly for the grass to remain acceptable to cows. It may be necessary to increase the ration given to the cows in the trough so that you can make more silage cuts.

Are there paddocks that you'd like to cut one more time in mid-October, at the end of the growing season? These paddocks should be cut or grazed in late August at the latest, even if there is only 2500 kg dm/ha herbage mass in the field. In late summer/autumn it will take another six weeks before you have a silage cut.

Ted clover as little as possible

Most of the nutrients in clover are in the leaves. When drying a crop, however, it's easy to lose them. The thing is to work carefully, preferably using a mower-conditioner and tedding the crop once. The next day, don't ted it, but swath it into rows and ensile or bale it. The crop from a field with a lot of red clover is best mown with a conditioner followed by forager. This aids the fermentation and prevents cows avoiding the coarse clover stems.

Cutting grass with blunt mower blades produces a frayed stubble and an irregular appearance. If not neatly cut, the grass will take longer to grow back, especially during dry weather. ▼

After its midsummer decline in growth during June and July, grass often gets a new impetus in August, which is why August is sometimes called the 'second month of May'. You can often produce high-quality silage in August. ▶

If you drive too fast while cutting, the cutting blades can't make a consistently clean cut. ▶

Manuring in late summer

Many fields do not need to be fertilised any more in August and September. If nitrogen is required, it will be in small applications. Providing an accurate application is not simple. It may be desirable to apply slurry onto some fields to supply potash and phosphate. This is particularly so where silage fields were cut often and have low to sufficient potash and phosphate levels.

Creeping bent grass responds more readily to phosphate than perennial ryegrass. Creeping bent grass with a purplish tint may point to a scanty phosphate supply, perhaps because of cold conditions. ▼

New grassland

Depending on the previous crop and slurry application, a small application of nitrogen fertiliser may be desired to get the grass off to a good start after re-seeding. About 30 kg N per ha is enough to harvest a light cut of 1000 to 1500 kg dm/ha.

More will be needed if you are sowing grass following arable, because its lower organic-matter content will provide less nitrogen than permanent pasture. Fertiliser advice for sowing grass after arable crops: in the first year, apply an additional 50 kg N per ha over the recommended amount. When sowing in summer, you can apply half the additional amount of nitrogen before the first cut.

Sandy soil vulnerable to leaching may quickly develop a potash deficiency. Clover is highly sensitive to a lack of potash. The first deficiency signs will be white or yellow spots on the older leaves. ▼

Reference of N supply

Perennial ryegrass is the best indicator of grassland nitrogen supply, and its colour is the principal clue. Other signals that may say something about the nitrogen situation include the development of the crop and the presence of crown rust.

1 *Nitrogen deficiency*

2 *Low nitrogen supply*

3 *Sufficient nitrogen supply*

4 *Excessive nitrogen supply*

Weed control

During late summer, the downward flow of sap dominates in weeds with strong roots, in particular docks. At this time of the year, it's worthwhile using chemical herbicides, since the poison will be distributed well throughout the plant and will also reach the roots. Late summer is the best time anyway to evaluate and if necessary kill weeds. This will prevent you having to spray again in the spring whilst the grass is already growing.

Grass-clover

Clover is sensitive to most herbicides, making it difficult to spray a grass-clover pasture. The selective herbicide Eagle (Bayer Crop Science) is very effective in grass-clover swards, particularly against dock. When you also have annual weeds in grass-clover such as chickweed and shepherd's purse, Basagran (BASF) can help. Eagle and Basagran can also be used in combination.

Tackling ragwort

If you only have a few plants of the poisonous ragwort, they can be removed individually. There are some useful chemical solutions on the market such as MCPA/MCPP, but this is only successful with young plants in the rosette phase. Spraying with MCPA/MCPP will still delay, or even stop, flowering and seed-formation in older plants. Cutting once does little to control ragwort, but regular cutting before the plant flowers will help to remove this weed.

Weeds with a strong root system, such as docks, are best destroyed with chemicals in the early autumn because of the dominant downward sap flow in this period. ▼

This new crop of grass badly needs topping. It's about to be choked by fat hen which is growing out of control. Cutting is enough to stop it. ▶

Chapter 7 October-November

Autumn turns to winter

In October grass growth slows considerably. Grazing winds down and the last silage cut is taken. The autumn is also the time to prepare the pasture for the coming winter months.

Carry out spot-checks to see if the soil has become compacted from machinery driven over it during the growing season. The amount of pressure needed to insert a soil probe in the ground is a measure of compaction. You can feel for yourself how easy or difficult it is for roots to penetrate the soil. ▼

▲ *The older animals can stay on the pasture until well into the autumn as long as there is enough tasty grass available. Keep a keen eye on intake of grass and the condition of the animals.*

Assessing quality

Because of the shortening of the days and drop in temperature, grass growth reduces sharply in the autumn. This makes planning pasture use less dynamic. Livestock gradually disappears from the pastures, you don't have to mow so often, and fertilising has already stopped.

Stay alert to signals from the grassland, even in the autumn. Conditions can change quickly at this time of year. Assess the quality of the pasture before winter sets in. Look at the sward density and the composition. A good tight sward offers more stability if punished with severe frost or rutting.

New grassland

If grassland has been sown in August-September, one frequently asked question is whether it's wise to take a silage cut before the winter. The answer will depend on several factors: for example, what is the carrying capacity of the soil, the dry-matter yield and amount/types of weeds? If you've sown around 1 September after forage maize, the grass will often be too short at the onset of winter. But with earlier sowing and using blends of fast-growing grass types such as hybrid ryegrass or festulolium (Westerwolds ryegrass), you can grow a good crop of grass on new grassland in the autumn in a short time. However, it must be cut for silage or grazed before the winter.

It's preferable to have newly sown grassland starting the winter short. This is more important with grass-clover than with pure grass. With too much grass, the clover will suffocate in the winter. If you can't cut in the autumn due to carrying capacity or weather, wait until frost sets in, even if it means cutting and ensiling in December!

New grassland is highly vulnerable to poaching. ▼

Puzzler

What is causing the poor grass cover here?

On many farms the paddocks nearest the farm have the poorest cover. The reason: the cows reach these fields first in the spring, and they graze these paddocks last before coming in for housing in the autumn. These paddocks are more often poached, causing good grasses to disappear and giving an opportunity for mediocre grasses to take hold.

◀ *Frequent and heavy autumn rains are bad for the soil structure. Waterlogged pastures are more vulnerable to all forms of pressure, such as traffic and poaching. So keep livestock and machines off them as much as possible.*

▲ *Autumn grass has a lower crude fibre content than spring or summer grass, so it breaks down quickly in the rumen. Respond to this by providing cows with more dietary roughage in the trough in the autumn. The high crude protein content of autumn grass allows it to combine well with high-energy products such as forage maize.*

Grazing during autumn

As the autumn progresses, the grass will be less readily eaten and its nutritional value will decrease so that fresh grass will make a decreasing contribution to dairy cows' diet. The decrease in intake and nutritional value is greatest on fields with a large proportion of 'moderate' grasses such as rough stalked meadow grass, marsh or bent foxtail or creeping bent grass. These and similar grasses quickly become less acceptable in the autumn and give the crop a musty odour. In the autumn, dairy cows must receive their base ration during housing. Anything they eat from the sward is a bonus. Grass intake in autumn is also highly dependent on the weather. In the autumn cows eat on average 0.75 kg dry matter per hour as opposed to 1.0 kg in spring and summer.

You can reduce the deterioration in acceptability and smell of autumn grass with good grassland management. Precede grazing with a silage cut as often as you can, and graze the sward with less herbage available. One disadvantage of keeping the cows on the pasture longer in the autumn is reduced nitrogen utilisation. With this in mind, you may decide to house the animals.

Don't graze grass older than a 28-day regrowth in the autumn, even if it has less than 2800 kg dm/ha herbage mass. Grass which is relatively short when you put the animals onto pasture will remain standing up straight. This will make it wind-dry sooner and it will taste better. ▼

In the autumn, the better acceptability of tetraploid grass often becomes even more apparent. ▼

Young animals longer outside

Yearlings can stay outside until well into the autumn. They have to be housed once the grass supply is insufficient, or when they start to poach the sward. Damage from poaching is the main reason for housing yearlings, but frost is also a consideration.

Hardcore tracks between paddocks will make them more accessible. Do take care that concrete tracks are free from loose stones. ▼

Poaching

Damage to the grass from poaching is also a risk associated with dairy cows in the autumn. Soil type and weather condition largely determine how susceptible the sward will be to poaching. You can minimise the risk with a few measures: don't graze more than fifteen cows per hectare in the autumn, and provide more drinking troughs in each paddock, at least one per twenty cows. Don't leave the cows too long in one paddock; give them a new one every day or so. Adjust the grazing time and paddock size according to the conditions. Assess the sward's carrying capacity and any poaching every day when you collect the cows.

Make the gateways to paddocks along the path wider than ten metres, to reduce the chances of poaching. ▼

Extended grazing: in the pasture all year

Extended grazing within the business system is a completely different approach to grassland use. Extended grazing is based on maximising the grassland contribution through grazing, as well as an optimum match of fodder quality to the needs of the cows. With extended grazing, the cows are on the pasture for as much of the year as possible. With a spring-calving herd, the quality and quantity of the grass keeps pace with the needs of the dairy cows.

To allow the best possible grass growth, extended grazing is based on turning the cows out to pasture with much taller grass covers than usual. The cows go in at 2700 to 3000 kg dm/ha and they come back out at 1400 to 1500 kg dm/ha of herbage mass. According to the extended grazing theory, long grass produces better than short grass. By keeping the ground covered with a good thick layer of grass, the pasture is less stressed by drought in summer.
To ensure that their cows can graze as late in the year as possible, extended grazing farmers strive for a supply of grass 'on the stem' in the autumn. This gives a different picture of a pasture than usual.

▲ *This machine is suitable for spreading straw over the land. Straw can be mixed in this manner with autumn grass to produce well fermented silage.*

Mowing during autumn

Mowing and making good silage is generally trickier in the autumn than during the rest of the growing season. The shorter days and often-changeable weather make it difficult to dry grass sufficiently within few days before ensiling it. Ensiling is already difficult at this time and is made even more so by the frequently high protein content of autumn grass and clover. The result is that grass ensiled in autumn often disappoints in nutritional value and intake.

It's actually best to dry autumn grass to silage with a dry-matter percentage between 40 and 50. Cows eat dry autumn silage better.

The fermentation of autumn grass can be improved by ensiling the grass along with a dry product such as dry pulp, cereal straw or grass-seed hay. You'll produce readily acceptable silage with good nutritional value. Farms that only harvest the cobs of maize can use maize straw, which combines very well with moist, high-protein autumn grass.

Avoid cutting too late in the autumn. Give grass plants time to re-grow a little, so the grass can start the winter in an adequate rest state. ▶

Autumn grass-clover harvest

The autumn harvest of grass-clover is often more valuable than grass alone. Compared with the beginning of the growing season, grass-clover has a high level of dry-matter and protein production in autumn. It's important to get the most from protein production of grass-clover, especially on organic farms, where protein often constitutes a small part in the diet.

One option which carries less of a weather risk is to artificially dry grass-clover at a grass-dryer. In fact you're making a concentrate. Drying changes the degradability of the protein, resulting in a product with higher undegradable digestible protein (UDP) content than the grass-clover you're removing from the land. With the grass pellets you've made, you'll save on the concentrate expense. Again, this is particularly attractive for organic farms due to the high protein content of grass-clover and the savings on expensive concentrates.

By baling autumn grass you can combine it easily with grass harvested in other parts of the growing season. ▼

▲ *Autumn grass-clover is suitable for drying into pellets. You have no more problems with ensiling and roughage is changed into high-protein concentrate.*

Puzzler

Take a different look

Take an alternative look at your pasture. The Google Earth programme lets you look at satellite pictures of your own grassland. The photo of this field shows clearly which parts have had machines on them, and where the chances of compaction are greatest. More information about grassland will be coming available via satellites in the next few years.

▲ *If you let grass-clover go into the winter with too heavy a crop, you run the risk that all the clover will have disappeared by spring.*

Preparing for the winter

It's economically attractive to use the grassland production for as much of the year as possible. In the autumn this means that management is focused mainly on keeping the grass tasty and attractive to the animals. Regular mowing and removal is necessary to provide the cows with good pasture grass. In autumn, cows prefer re-grown aftermaths.

Work towards good grassland quality so that the sward starts the winter in good shape. A height of 7 to 8 cm gives the grass the best chance of good survival through the winter. If there are still tufts after the last grazing and the grass isn't evenly short, go over it again with the topper.

Red clover contains phyto-oestrogen, which resembles the human and animal hormone oestrogen. Sheep in particular may experience infertility when eating red clover during the breeding period. ▶

Winter grazing with sheep is a good way to fight unwanted ragwort. Sheep eat the new seedlings and young rosettes, stopping the ragwort from spreading further. Since enzymes in their digestive tracts neutralise the toxic alkaloids in ragwort, sheep are less sensitive to this weed. ▼

Looking back

The end of the grazing season is a good time to evaluate the year's grassland use. To do this properly you should have a lot of information stored using a pasture calendar or computer programme. You can learn a lot from all of these checks.

Check to see if the pasture is ready for the onset of winter. Don't look only at the sward, but also check furrows, ditches and drains. ▼

Look at the following when reviewing:

- Make a list of usage, by paddock. Evaluate whether you've succeeded in varying cutting and grazing on mixed-use paddocks.
- Determine the total dry-matter yield per hectare for the growing season just ending.
- Compare the different fields with each other. Look for possible explanations if yields of certain fields are lower than the average for the farm.
- Evaluate fertilizer use. Compare the amounts of nitrogen, phosphate and potash applied with the yields achieved. Test whether the pasture was given as much fertiliser as it needed, and whether slurry and chemical fertilisers were evenly distributed over the paddocks. A few paddocks may need improvement next year, perhaps in the spring.

▲ *Topping is a practical means of preparing the pasture for winter. Choosing the right time is not always easy. There cannot be any chance of night frosts, and often in the autumn the pasture's carrying capacity leaves much to be desired. If there is rutting, the disadvantages outweigh the benefits.*

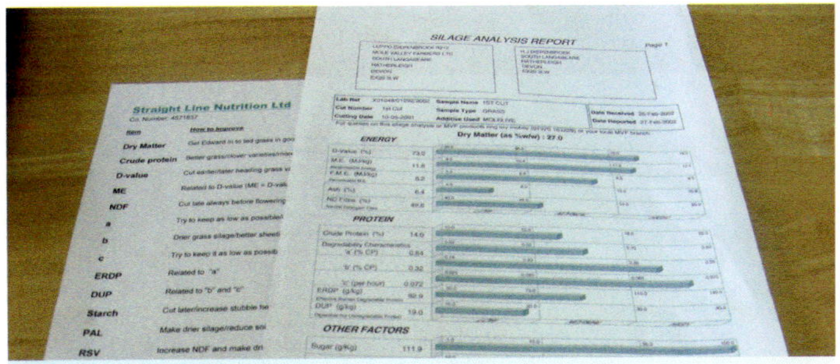

Feed analysis

Autumn is a good time to evaluate the grassland crop using the results of feed analyses. Look at nutritional value and digestibility as well as how well fermented it is, which you can obtain from the ammonia percentage. Crude protein gives an indication of fertiliser use and uptake, and crude ash of contamination. Weigh up the weather and seasonal influences along with the figures when you do this evaluation.

Music to your ears?

Many arable farmers are already using GPS and infrared images, but this is worthwhile for livestock farmers as well. Infrared photos taken from a plane can measure crop densities to within half a meter, which is then converted to the relative biomass within the field.

Differences can be accounted for by differences in fertilisation, drainage, soil fertility, root growth, etc. Using the chart, you can dig profile pits or take soil samples and get an idea of any problems. Entering a final fertilisation chart in a GPS system on a fertiliser spreader enables you to fertilise exactly where you need it for a maximum and uniform grass cut. This system will also indicate potential losses from such points as waterlogged areas. If the loss is five times what it looked like to you, it may stir you into action.

Photos and maps can be useful for 3-4 years (as long as a soil samples). And it's not that expensive. Accurate infrared photos from satellites are not yet possible, but this is a matter of time.

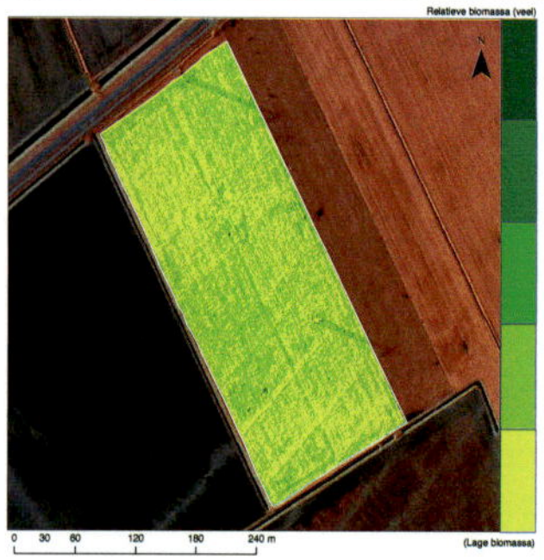

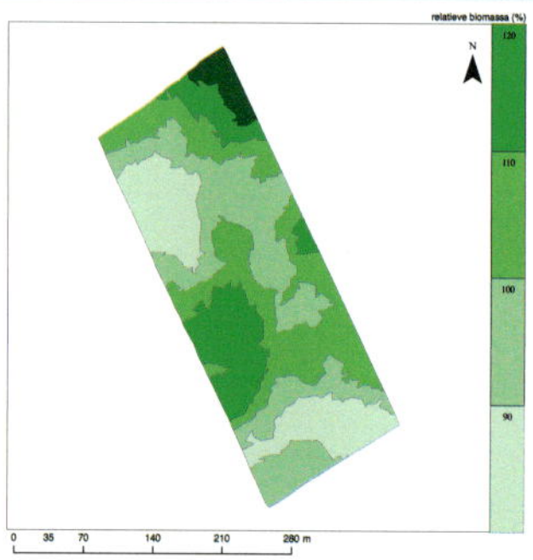

Field image produced with an infrared camera. The same field with biomass calculated.

Handy aids

The best known aid for optimum pasture use is the land-use calendar: originally on paper, now also integrated into management software. A few more resources:

Guides

The Wageningen University Animal Sciences Group has several guides, available at www.asg.wur.nl (in Dutch only): Producten > Veehouderij > Software.

BeweidingsWijzer: calculates the results of changes to the grazing systems (labour, economics, manure).

HerinzaaiWijzer: determines if improving a sod is worthwhile. Looks at costs and anticipated extra yield in the first five years after re-seeding.

GraslandGebruiksWijzer: makes a calendar for grazing, growing fodder and fertilising grassland and calculates production, quality and losses in grazing, cutting and ensiling.

Other interesting links

British Grassland Society — www.britishgrassland.com
Local grassland societies as well as national/international events based around grassland.

DairyCo — www.dairyco.org.uk
Grassland management for milk producers

Kingshay Farming and Conservation Ltd — www.kingshay.com
Independent Soil, silage analysis, advice and costings.

National Institute of Agricultural Botany (NIAB) — www.niab.com
Independent UK plant testing institute

Department of Environment, food and Rural affairs — www.defra.gov.uk
Search for RB209: fertiliser advice and nutrient book RB209

Some grass seed suppliers

Barenbrug — www.barenbrug.com
Eurograss — www.eurograss.com
Limagrain Advanta — www.limagrain-europe.com/gb

Index